Fruitcake Hill and Beyond

A Sequel to Fruitcake Hill

Gerald J. Kuecher, Ph.D.

CCB Publishing
British Columbia, Canada

Fruitcake Hill and Beyond: A Sequel to Fruitcake Hill

Copyright ©2013 by Gerald J. Kuecher, Ph.D.
ISBN-13   978-1-77143-089-0
First Edition

Library and Archives Canada Cataloguing in Publication
Kuecher, Gerald J., 1951-, author
Fruitcake Hill and beyond : a sequel to Fruitcake Hill
/ by Gerald J. Kuecher. -- First edition.
Issued in print and electronic formats.
ISBN 978-1-77143-089-0 (pbk.).--ISBN 978-1-77143-090-6 (pdf)
Additional cataloguing data available from Library and Archives Canada

Cover artwork modified from free WORD clip art by Allison Taylor, artist.

The recollections presented in this document are largely those of the author. The author apologizes in advance if misrepresentations, inaccuracies, or omissions are encountered or perceived. Sibling names were not directly cited in this manuscript to protect their identities.
Genealogical information presented herein is the product of a multi-year search.

Extreme care has been taken by the author to ensure that all information presented in this book is accurate and up to date at the time of publishing. Neither the author nor the publisher can be held responsible for any errors or omissions. Additionally, neither is any liability assumed for damages resulting from the use of the information contained herein.

Publisher: CCB Publishing
              British Columbia, Canada
              www.ccbpublishing.com

Acknowledgments

This book is dedicated to my parents, Bob and Babe Kuecher and all the interesting characters and situations I have encountered at Fruitcake Hill and in my life's journeys away from "the hill."

I wish to thank my wife, Jean, my older sister, Julie, and my youngest brother, Ed for their un-ending support of my so-called creative endeavors. I also recognize neighbors and friends Hot Dogs B., John R., Jeff B., Jerry B., Kevin and Sheila M., Lori H., Athens S. and various Kuecher siblings for their encouragement to document this family history from Palos Township, Illinois. In addition I thank numerous people I met along the way in the oil business and in academia as well as the characters and friends of the family mentioned in this book and those not mentioned, yet equally deserving.

This book consists largely of the personal reflections, observations and opinions of the author. Various siblings and colleagues were asked to review the work prior to publication. In this way the veracity of the author's stories were corroborated.

The author made considerable efforts to mask the identities of characters and companies in this book. No harm was intended towards any one or any company in writing this book.

Prelude

Fruitcake Hill, published in 2008, lays the background for this story. This document can be read independently, however, if a brief introduction is provided, courtesy to the newcomers.

My family lived in an ancient white farmhouse built in 1871 atop a hill some 23 road miles southwest of the rapidly expanding City of Chicago. The house nearly burned to the ground in 1928 when a 3.5 year old toddler, who later would become our mother (Babe), got into mischief at the cook stove when left alone in the house. Fortunately, fire trucks were able to break through a layer of ice in a nearby gravel pit and pump water to the fire. Babe survived, but was haunted for years about her role in the conflagration.

Our farmhouse was located on a hill in the middle of a very large turf nursery. The owner of the turf nursery purchased the land (76.62 acres) from Babe's mother, Florence in 1947 and the monies obtained in that transaction supplemented her earnings as a teacher in the City of Chicago.

Babe spent her pre-school years on the farm but Babe's sister, Mary, was raised largely in the City of Chicago. These sisters were close friends but became very different people because they were raised so differently. It was a case of country mouse and city mouse, despite the fact that both were educated at the same school (Academy of Our Lady) in the south Chicago suburb of Beverly.

Both Babe and her sister, Mary, were courted for marriage by the man who would become our father, Robert Kuecher (Bob). In fact, the courting became sufficiently entangled that when Bob asked Babe's mother for the hand of her daughter in marriage, Babe's mother asked, "Which one?"

My mother, Babe, was given the farm house and 3.38 acres of property by her mother, Florence O'Connell in 1955. And on this property, Bob and Babe raised a family that ultimately numbered fifteen (i.e. thirteen children and two parents).

The house was very old, contained little insulation, and had only three bedrooms, one of which was our parent's room. Relief would come in 1967 when Bob and Babe built a new addition on the old house. Instantly the bedrooms increased from three to six and bathrooms increased from one to three. This provided some privacy for our family and its visitors. Yet the kids pine over the idea of the tightly packed community we once had, and perhaps how the family compromised some of its closeness when the new addition was built. The farmhouse and the hill, however, only frame the story of the people that lived here.

Since my book *Fruitcake Hill* was published in 2008, I have been encouraged to write again. This book, *Fruitcake Hill and Beyond (2013)* follows one of the family members (the author) from that ancient farmhouse and his professional journey to adulthood. And now with a family of his own, he wonders how Bob and Babe raised such a large family in such a small living space. These questions, and more, challenged the author's ability to tell the story as it really happened.

Fruitcake Hill and this current work are compelling reads because the characters introduced along the journey were so colorful. In that sense, the author did not need to create the characters. He needed only to report on them. The stories that follow provide some insight into the author's life's journey and partially represents what newscaster, Paul Harvey, would say was "the rest of the story."

Gerald J. Kuecher

Fruitcake Hill and Beyond

I grew up in a very special place, I had very special siblings and parents, and I had a happy childhood. Those, at least, were my working hypotheses growing up in Palos Township, Illinois.

But life wasn't perfect. And I now wonder if my mind somehow remembered things differently than they actually occurred. If this is true, there may be little hope of getting at the real truth. Perhaps that's good. A more balanced assessment is to consider us all wounded warriors of our childhood, i.e. still able to love, but also capable of uncharacteristic behavior when those bad memories raise their ugly heads. And life on our hilltop was no exception.

What I have come to believe may be much more important than who I was. I believe I should expect miracles. This positive thinking comes true more often than not in my life experience. And even if a plan or event crashes, I rewrite the event to be positive. This is how I operate.

"Expect to be surprised" was the final advice given in the movie *Dan in Real Life*. I look every day for my surprises and I am convinced my life consists of more surprises than the life of the average Joe. This does not mean my outcomes are better, only that the process may be more intriguing, or at least I think so.

Reflections

I am sitting on a chair in my back yard and my wife is cutting my hair, what's left of it. I have a towel around my neck and it's hard to imagine just who or where I am. I feel stunned and uncertain. I look for clues in my back yard and my wife re-positions my head with each move. So instead of scanning my environment, I focus on the pink bougainvillea plant diagonally across the yard that is athletically scaling the wooden fence. And here I remember my dear Fluffy, the beautiful Samoyed mix dog that was buried in front of the

3

plant's main trunk. And off to the right my eyes are drawn to the burial site of our special cat, Kassy buried about three feet away. And my world slowly comes into focus while I adjust to my jet lag. I am slowly de-programming from my life in a foreign land and adjusting to life at home again.

My disorientation extends to waking in the night. Am I in my home bed or am I elsewhere? The room is dark so which way should I turn to head to the bathroom? I gamble on going to the left and I walk into the wall. Wrong again! This current adjustment results from a stay of over four years in Saudi Arabia. But this has happened before. I once knew these environs and now all this seems so unfamiliar. Time has a hand in these things.

It's like that when I return home. I see my world only occasionally, noticing mainly the changes, not what has remained constant. It's like reading only the first and last page of a book and wondering why everyone thinks the book is so great. Then I think of my Dad, my brother John, my cousin Dorothy, my aunt Mame, Dick Cronin, Grandma Kuecher, Nana, Aunt Gladys and all the faithful departed of our family. And although I am sorry they have passed from us, I am nevertheless damned glad I am still "this side of the grass."

So here I am, licking my wounds. Most of my wounds resulted from isolation. And although I had many friends in Saudi Arabia, Indonesia, Egypt, Kazakhstan, and elsewhere, my wife and family were often not with me. So I paid the price. "The family needs shoes" was my common defense. And if you need money, living full time overseas is the ticket. Workers get big premiums for the inconvenience but there are just some things money cannot buy.

A lot has changed, for sure, and I feel like I was dropped from a car on a high speed expressway. The world is turning yet I wish for it to slow or stop to get a good look. I need to adjust but the jet lag and cultural abyss has taken its toll. Yet we itinerant geoscientists continue to travel to distant lands for the opportunity of seeing remarkable things and making good money. Living in a foreign land is a unique experience.

This is not my first trip around the block. I have had extensive stays in a number of foreign and domestic locations. But let's just say my journey has been quite different from the average Joe's.

My wife says there will be a major adjustment when I come home to stay because in previous visits, activities focused on me. I had doctor and dental visits and I shopped for clothes, shoes and electronics. But when I am home for good, I must be the one who does the giving. That's fair, I guess. And my wife says she is making a list of behaviors I need to improve when I am again a permanent resident.

So as I awake to my new world, the stage is set to share my journey from my life on an Illinois homestead in a Catholic family of 15 through my years as a professional geologist and on to my retirement years. With that, let's begin.

Special Memories, Special People

The Early Years

I was the seventh child of Babe and Bob Kuecher and my family called me Tiger. I was born in 1951. Babe (my mother) went on to have six more children, thus I was ultimately the middle child with six above and six below me in age. Years later, I asked my mother just why she had 13 children and she responded in her inimitable Irish logic that had she stopped at six, she never would have had me.

The Kuecher children were rather proud of our family size and treated it like a local Guinness record. But the facts indicated we had some rarefied company in nearby communities, e.g. the Johnsons and the Diamonds reportedly had 14, the Barbush family had 10, the Warringtons had 7, the Lucas family had 9, and the O'Connors had 8. Large families got special recognition in those days. We received special privileges at the supermarket, at the school, and at church. And regarding tardiness, who in their right mind really expected us to be on time? I remember most clearly our walk past those already seated in the elementary school. It was *heads down* from the front of the room to the back hoping against hope no one would notice. But of course they did.

I remember getting on the bus for the first time to go to school and sitting with another terrified first grader. I felt a part of me died that day. It was like being kicked out of my own house. Hey, just because all my brothers and sisters before me went to school did not make a convincing argument why I should. I felt just fine being at home and did not understand why I must be weaned of my home respite. My older brothers and sisters noticed my trepidation and prepped me to be brave. That seemed to work.

The biggest event in our elementary school lives was First Holy

Communion. Preparation for First Holy Communion took forever. First we had to undergo extensive catechism classes and sit patiently while Father Coy and Father Bron lectured us on how this was going to go down. We were not asked to provide advice. Instead we were positioned in rows and told to approach the altar with mouths closed and hands clasped. The Communion photo, taken some 50 years ago, cannot hide the fear in the faces of my classmates, however. "Mess up and you're dead" came across quite strongly.

Aunt Glad, my father's sister, cooked for the nuns at our school and was dearly loved until her death in 1976. Glad was well known for her culinary skill and that resulted in my box lunch regularly fetching twice the price of the runners-up at the charity box lunch auction. The poor kids offering those top prices had no idea that Babe (not Glad) made the lunch. For those prices they could have had lunch at Marshall Fields. Instead they received more conventional fare, like grilled cheese sandwiches and Twinkies. This was my first experience with "Let the buyer beware!"

Many of the kids had extra-curricular activities and stayed after school. Of course that laid a burden on Babe as she would have to drive to school and retrieve us. But Babe was seldom on time. So we waited, and we waited, and we waited. That was the high personal cost of extra-curricular activities in a big family. And because nobody then had cell phones, Babe may arrive at the pharmacy pick up location hours later to find we had already taken a ride with someone else. So in our effort to pass the time, we often sang a spirited round of *99 bottles of beer on the wall*, a song locally referred to as the Irish National Anthem.

My grandmother (mother's side) once took me for a ride downtown Chicago. I had never seen the downtown area and was quite excited to see the 41 story Prudential Building, the tallest building in the city at that time. I remember standing on the curb and looking up, and it appeared the building leaned over the street. And when we entered the observation deck and looked out at the buildings below us, I suddenly had little faith in the contractors who built it. Many years later I worked on the 47th floor of the 80 story Standard Oil Building and my office looked down on the observation deck of the Prudential Building. How times do change!

Babe meanwhile faced the daily tasks of shopping with unwilling children in tow. There just seemed to be nothing more boring than shopping for someone else. But Babe, in her brilliance, realized she could drop the boys near the monkey cage at the mall and sure enough, they could be found at this location an hour or more afterwards. We were connecting at a primal level.

Regarding recreation, the boys uniformly shared a passion for hunting. If the boys had a box of shells and a shotgun they could hunt with their friends and brothers. This was rite of passage stuff. Older boys would give hints to the younger boys to keep in lines and not let the dogs run too far ahead. A simple "yip" or "watch" or "coming your way" were code words we all knew to hunt together. And if a younger boy downed a bird the event was widely celebrated by the other boys, like a lioness making her first kill. We became men in those fields. My father was probably the catalyst of that hunter mentality because he too was a hunter. Moreover, game killed during the week meant my Dad's mother would come out on the weekend to cook for him. The worst part of eating the wild game we shot, however, was biting down on the buckshot.

Granny D

When I was young, perhaps in sixth grade, I began hanging out at the home of Granny D, the grandmother of Hot Dogs and Hog Jaw B. Granny came to the US from Ireland as a 17-year old and took a job as maid for Chicago attorney Clarence Darrow. But apparently Granny never became a citizen. How that happened we don't know but she voted every year. Go figure! In 1968 she went to vote at a polling station and the clerk asked if she was an American citizen. Granny bristled and said "I've been voting as long as you've been jumping over the hill." Of course, voting irregularities are no stranger to Chicago where they advise voters to come early and vote often. And in Chicago, Granny fit right in.

Granny was an ancient old coot who wore a very simple, faded pull-over dress and raggy old socks. On cold days she wore a sweater. Granny had just a few remaining teeth and sported a few long gray hairs on her chin. Occasionally Granny fell asleep in her chair and snored, and occasionally she would stir to whack a fly that was bothering her.

Granny lived in a small two-story home that had a steep stairway and she had a cat with which she had a fairly acrimonious relationship. One day, Granny noticed her cat had eaten some food she prepared for herself so she followed the cat to her upstairs bedroom. And noticing the window was open, Granny took the opportunity to launch the cat out the second story window. Take that!

To get to Granny's house or to get to the tavern to buy soda, we had to pass a yippy black terrier owned by our neighbor George M. This dog chased us nearly every day. Our strategy was to gain sufficient speed on our bicycles so we could cruise with our feet up and away from the dog's nippy jaws. My older brother Lui just kicked him.

Granny couldn't walk very well so when we knocked on the door, it usually took her some time to get there and ask in her creaky voice, "Who is it?" One day, in fact, two of us were being pursued (for hunting) by Palos Hills Police Sergeant F and we made a bee line to Granny's. But by the time Granny, in her arthritic condition, could open the door, Sergeant F saw us enter Granny's place. We sat down in the kitchen, hidden from view and sure enough, Sergeant F pulled his squad car into the driveway, walked up to the door and knocked. Big mistake! Granny picked up her broom, opened the door and came out smoking. She beat on Sergeant F, whacking him several times. We, of course were laughing uncontrollably from our kitchen vantage. You don't mess with Granny or her boys. I never met a person so universally in your corner as Granny D. She had the loyalty of a pit bull.

Granny seemed to favor us over her own grandchildren when it came to cashing in on returnable bottles. She would make secret arrangements with us to come with our wagons when her blood kin were not home, netting my brother Billy and me four or five dollars a crack. That was big money in those days. We helped Granny by getting that mess off her porch and she helped us with the cash.

One day, Granny said she was going to make us a batch of blueberry muffins. So my brother Billy and I resigned to the TV room while Granny began her cooking magic. About an hour later, Granny shuffled her way into the TV room with a batch of blonde muffins. "What's the deal, Granny?" we asked. And Granny whispered apologetically that she forgotten to add the blueberries. Granny did her best but she was quite old and forgetful. So we laughed and decided to slice the muffins and add the blueberries manually. That was fine by us. Granny was so cute.

Granny had a bit of gambler in her personality. She regularly attended BINGO games at the church and loved scratch-off and bottle top games as well. Granny was arrested some years later in Summit, Illinois attending a BINGO game with other senior citizens and her name was posted in the local paper. I would love to have been a fly on the wall that evening.

When Granny would catch a cold we would tease her saying "We may have to shoot you Granny." At which Granny would break in a grin and flick her hand out as if to say, "Oh go on now." Granny D died in 1968, a long-time resident of the U.S. and citizen of Ireland. What a character and what a friend!

Bri, Mott, and Shonitz

In eighth grade, perhaps 1965, my elementary school played the Oak Ridge public school basketball team and I ran into a fellow much like me. His name was Bri (sounds like rye). I saw in him a competitive fire and knew I would make every effort to be friends with him. He and I, although presently representing different schools, would attend the new high school the following year.

I called Bri on the telephone several weeks before high school commenced and arranged an overnight. I set up my Dad's tent in the back yard and all was set up for a memorable evening. I even had a goose cooking on hot coals underground that I shot the day before. And with all the excitement, Bri was ecstatic. He was a city kid and did not know much about country living. And despite living on the south side of the City of Chicago, he became a fan of the Green Bay (Wisconsin) Packers.

All my fears of Bri evaporated on our first meeting. He loved my home, my siblings, and the outdoor environment, and we talked into the night. At one point that evening, I realized I had forgotten a frying pan to cook by the fire and at approximately 2AM I asked Bri to go into the house and get one. So Bri swaggered his way into the house just after my big brother John returned home from a night out. John saw this stranger walking up the steps and brazenly open the door. John did not know what to make of this and stood behind the door poised to pounce on him. Instead Bri, upon making eye contact peeped, "Gotta frying pan?" We entered a lifelong friendship that very night.

A few weeks later, Bri and I took a shotgun and went down by the creek to shoot blackbirds. I had been talking up my shooting prowess and it came back to bite me. Bri and I were sitting in wait for blackbirds when one flew directly over us. I took aim and *blam* it fell from the sky and crashed into the creek below us. "Did you see that? "I exclaimed! "I knocked the hell outta that bird" I proudly said. And no sooner did I say that when Bri looked down and saw the bird swimming and then flying away. He loved it when I embarrassed myself and never forgot such moments. This is what made us such good friends. We laughed at our vulnerabilities and that made the images we created of ourselves fall apart in such a human way.

I wonder, occasionally, just what I saw in Bri. He was an average student. He got in trouble more than the average Joe, and lived in a home with his mother and her "friend" Fred. One could tell there was tension whenever Bri and this "friend" were in the same room. Something was amiss and maybe this is why he was troubled. Fred tried to discipline Bri but had no authority because he was not Bri's father.

Bri and I were inseparable friends during the first two years of high school. It was about this time in high school that I recognized my calling in the physical sciences. Bri, Mott (a neighbor and friend of Bri's) and I were in the same class of Earth Science. And we studied together, largely because they didn't get it. Bri and Mott called the teacher "Otto" referring to football player and coach Otto Graham. And we named our softball team the Piedmonts referring to a foothill plateau region in the Appalachians. Otto freaked out one day when our whole team wore Piedmont shirts to school. I guess he thought we were some sort of gang.

Bri was plagued by lousy, bullying gym coaches that spewed demeaning nonsense. Bri could have believed them when told he couldn't do this or that. And one may wonder about the outcome of Bri's life had he not been strong enough to stand up to that criticism. Who could ever forget the day Bri quit the football team and ran by our after-school practice in his new cross-country outfit. The coaches stood there aghast! That was an in-your-face Bri moment.

Bri was an incurable romantic and listened to Johnny Mathis albums well into the night. Bri, in fact, could sing a good rendition of

Johnny Mathis and would break into song for no good reason. He was living and dreaming of a better day and had a Don Quixote vision of the future. Bri's mother would make us all a nice breakfast when we rose on non-school days and I felt nearly as much a part of his family as my own.

Occasionally Bri and I would share a humorous story of an awkward encounter with a female classmate. We were immature, for sure, but we celebrated our immaturity and poked fun at each other for failing in embarrassing situations. One recurring theme was the girl we just could not muster the courage to ask for a date. And God forbid, if she did consent, how should we contain the moment without acting incredibly uncool.

Bri was an excellent pitcher in the Pony Leagues and showed me all his trophies. I was happy for him and it was true, he had an excellent fast ball. But I knew he had not yet seen the way I could throw a ball. He learned during a game of high school bombardment. In our first game I knocked a prisoner unconscious, wrapped the impression of the ball around Bri's leg, and opened the gym doors with my fast balls. Bri, however, was man enough to give people around him credit. This was one of his greatest character assets.

Most would have thrown in the towel on Bri in high school but he was an incredible competitor. And I knew he would land on his feet despite his poor start. He played pool to win and did some boxing as well. His bad boy image was complete with his pompadour hair style, Cuban heel shoes, and skin-tight black pants. To be truthful, there were times I looked at him and muttered "smart ass" in desperation.

Both Bri and I remember one defining moment that we both consider the greatest moment in our lives. It occurred at a party in my parent's basement when we were high school juniors. Lots of young and middle age people were there, the beer was flowing and the music was a heavy riff from the album Cold Blood. For some reason a circle formed and people took turns dancing in and out. It was spontaneous and perfect. Bri for his turn entered the circle doing a chicken dance and shuffled his way out on his Cuban heels. The amazing thing was not a single hair on his head was out of place. Well how could it with all that hairspray? We called it our *crystal moment.*

Bri had another sidekick that lived nearby. We called him Shonitz. Shonitz would do most anything for a laugh and we would do most anything to him for a laugh. There were few boundaries on our behavior. I remember one day we were all headed to the Chicago Auto Show and Shonitz was bragging to Bri and Mott that he didn't need to worry about me. He said "Kuech won't hit me." And at the third time I turned around and whacked him right in the mouth. The entire car erupted in laughter. So our standing joke whenever Shonitz and I got together was "Kuech won't hit me."

Bri, Mott, and I would enter into well-timed vignettes with Shonitz and they went something like this: We would approach Shonitz and stare at him. After a time Bri might say "You did, didn't you? "Do what?" Shonitz would reply. Then I'd pitch in "Yep, he did." And after a few moments Mott would say "He did, didn't he!" Poor Shonitz was totally outgunned. Another time Bri, Shonitz and I were in Brian's basement playing pool and waiting for my sister to drive us to the movie. When we heard the horn outside, Bri ran upstairs. I followed but brought a boxing glove and when Shonitz turned the corner he ran right into a punch and was on his back and out like a light. Girls probably would not accept this behavior but we were guys and to us, this was hilarious.

Shonitz' mother was a real nut case. She was Croatian and referred to me at times as "that German sob" despite the fact I saved her son from his own destruction on numerous occasions. I'll never forget the flashlight flying through the air at me when I returned Shonitz from a drinking binge. Bri knew Shonitz' mother would be waiting up for him and let me take him to the door. And of course,

Bri was laughing uncontrollable in the safety of the car. Years later I am reminded of that night when I open Facebook and it asks "Can you recommend a friend for Bri?" and I am tempted to suggest Shonitz' mother.

Shonitz and Bri once made me a proposal to get my scooter running. We negotiated on a rate and I agreed to pay them 80 dollars for repairs. They worked two weekends on the project with Bri issuing orders I'm sure because he knew nothing about repair. Then one Saturday afternoon they called and said they were coming over with the scooter. I was so excited seeing the scooter with these two jokers coming up the hill and into our yard. Bri was smiling ear to ear with joy but after just two turns around the circle they said they still needed to make some adjustments so off they went again. I can still see Bri hanging off the back of the scooter laughing like there was no tomorrow. Well the sad thing for me was that they never again got the scooter running. So I was out 80 dollars and all that was left was a junked scooter. Afterward Bri would shriek Scooter!" in falsetto whenever he saw me. You know, I think he owes me 80 dollars adjusted to today's cost of living over a term of maybe 45 years. The next time I see him I'll make that pitch!

Teenage boys have a thing about competition and that is no doubt why high school sports are so popular. One of our favorite ways of testing our strength was with arm wrestling contests. One evening that competition led to an unforeseen accident. About ten senior high school boys had placed orders for drinks at a local pub (we had phony ID's) and were waiting to be served. Bri challenged me to arm wrestle him and after a good tussle, I won the match. But in my elation I stood up just as the drinks were being delivered over my shoulder and bang, the drinks flew everywhere. We were asked to leave.

Shonitz died a few years later in a bar from a heart attack brought about in a spirited arm wrestling match. Mott remarked "Dying in a bar-one couldn't have written Shonitz a better epitaph."

In retrospect I am sorry for the way we treated Shonitz. We contributed to Shonitz propensity for risk and his ember burned out early, a lot like his look-alike John Belushi.

Bri and his mother moved out of the Palos area in our junior year of high school and went to live with his brother Jerry in Seabrook, Texas. I don't know if Fred was asked to come or told to stay home but he died soon thereafter. I wrote to Bri and called him occasionally. He knew I cared and that was the most important message. We would get together again after he finished high school in Texas.

Bri joined the US Navy after completing high school and served 1969-1973. I think the Navy was good for him. It got him off the streets where, knowing him, he would have been in trouble. But just how would Bri go about getting the job that would use his talents. It was certain his path would have to take him through college. But he hated high school and I thought he would hate college as well. But Bri learned discipline in the Navy and later graduated with a BA from Roosevelt University in Chicago.

And out of the ashes, a new man emerged who would master computer sciences, or at least the vocabulary to manage computer operations. Today Bri is a Senior Vice President for a large health insurance firm in Chicago with a staff of over 2,000 employees. Who would have figured as he whined *Kuecher* in falsetto into my English class at lunch time or roared with laughter when I shook peoples' hands with my pheasant claw that this would be a highly successful businessman! No one would have thought he could stick with anything, yet this year he is currently enjoying his 38th year with the company. They say "every saint has a past and every sinner has a future."

I have always been a great encourager of my friends and I wish them well. I am happy my friends are successful. I believe the Tiger of high school knew Bri had some really good things going for him

Gerald J. Kuecher

that virtually everyone else missed. And we had a great and lasting friendship. Reveling in others success is only half of life's story, however. We need to be thankful because the outcomes could have been different.

In all, Bri had two marriages and produced four children. I used to tease him that he knew how to marry women but not how to live with them.

At the Stagg H.S. 40th reunion, Bri was in his element. I was not there, but apparently Bri's French teacher was, and when she approached, Bri began issuing a litany of French phrases he had not used in 40 years. And after a minute the teacher said, "OK now go sit down!"

Bri met a sweetheart of a girl at the re-union. She was there alone and they exchanged email addresses. A few months later, both attended a picnic I staged in my parent's back yard. It was good to see both of them happy with each other. She had considerable misfortune in her own family. She lost her husband suddenly to a mitral valve failure and lost her son to a rare disease. That's more heartache than a mother her age should have to endure.

Bri is a Swede by origin and they tend to live long lives. I recently saw a picture of Bri sitting on his desk at work and he looks remarkably young. He'll probably outlive us all.

Shugs

My friend at AA Stagg, once Bri left for Texas, was known as Shugs or Sugar Bear. Shugs was a wrestler in high school and worked a chicken fry business by night. The manager of the chicken fry was a man named Sam and he interested Shugs in campaigning for Sam's political office. It didn't seem there was anything in it for Shugs, however. My friend Herb later joined the political world and both he and his wife ultimately had some success in politics. They

17

had associations with Sam as well.

Shugs had a high school girlfriend. He was madly in love with her and she was only mildly in love with him. And when she finally left him, he sunk into a severe depression and drank himself to stupors. In his first month following the break-up, he set a mark to get drunk every day for 30 days. We laughed and asked, "Hey Shugs, what day are you on?" but in fact we were witnesses to an unfolding tragedy.

Shugs later invested in a bar up Kean Avenue and that is always a poor idea to have an alcoholic as bartender. Lessening his chances at a long life, Shugs also was a heavy smoker. Many years later I heard Shugs had cancer and I flew home from Houston to see him. It was really sad. He was lying in a medical bed in the front parlor of his home and he was a mere shell of what he used to be. I remember my inadequacy in talking to him about his condition. And it became clear there is a great conspiracy among the living to guard the fact someone is dying. "Hey Shugs you're looking great" and other lies were bantered about but he knew and I knew he would not survive this. He looked like hell. And when I left he said "good bye" in a very profound way. "See ya later Shugs" and other nonsense was issued by me as I backed out of the house. I didn't have the courage to say good bye myself. He died in 2001 at the age of 50. Our birthdays were only five days apart. He was special. I want to issue a hearty thank you to Mike and Herb for taking care of Shugs in his final days while I was in Houston. They were great and loyal friends.

A philosopher once maintained we may experience more misery than happiness in our lives and added the painful experiences were likely more intense than the pleasurable ones. Case in point e.g. an hour of toothaches may outweigh a week's holiday. This is a sad commentary when I consider Shugs' final days. It must have been hell. I pray for a merciful ending when it is my turn. I am happy I am still "on this side of the grass" however, and give thanks daily for another day.

JR

JR was a friend that lived at 111th Street and Kean Avenue. He called me Nipsy. JR and his family lived alongside cousins in a row of houses down by the canal. Many of the Palos families were involved in the canal in one way or another. Black dirt, sod, soybean, and corn farming industries were made possible by the canal's organic bottomlands.

JR's younger sister was a cutie and I had a crush on her. But I only had the nerve to ask her once on a date. Interestingly enough, my older brother John's first date was with JR's older sister. That's how it was in the country.

In elementary school, JR served as a patrol boy, stopping traffic both ways on 111th Street and allowing safe passage for young people at his corner. One day, despite JR's good work, a car sped through the stop and struck a cousin of JR's crossing that street. We prayed all day for that young boy and he did survive despite a broken skull, as I remember.

JR had close friends named Doug and Jake. Doug and Jake knew of a summer job that paid well at the Zion nuclear plant north of Chicago and asked JR to come help them. But JR could not get to work until the following day because he had to first resign from his plumbing job in Palos. This delay getting him to the job site likely saved his life because Doug was killed and Jake terribly burned that very day in an industrial accident at the plant. The boys were cleaning a large culvert with solvents and welders working above them ignited a flash fire below. JR certainly would have also been injured or killed that day had he not been saved by good fortune.

JR and I once took a neighborhood kid hunting. He was a wrestler on the Stagg team and his name was Mark. I remember giving him detailed firearm instructions in the hope he would not kill himself or kill us. But some lessons are better learned in the field and thank God the following mistake was not fatal. Mark placed his shotgun against a chain link fence and began scaling it. But he forgot to leave the safety on and I heard a loud report. The blast went off maybe a foot from JR's chin. And after I dressed this young man down for his mistake, I silently gave thanks to God for escaping another of life's close calls.

JR went to Stagg HS and remembers a funny incident that occurred at a Stagg HS football game when I was on the team. Stagg was getting killed and the other team was running up the score. So JR's dad started a chant "We want Tiger, We want Tiger." So the football coach found me along the sidelines and said, "OK, go in there and do something." So I ran to the huddle and told the quarterback "You're outta here." He was shocked, but left. And I called a play in the huddle, telling my eligible receivers to "go deep." I don't remember the outcome of the play but the act of telling the quarterback to get out was legendary. Apparently the coach intended I go in the ballgame as a guard (my regular position), not as the quarterback. But he told me to go in and *do something*. Now I ask just how I was supposed to do something significant from my guard position? I only played one play as QB but it was funny beyond belief. The coach went bonkers.

JR later attended Bradley University in Peoria, Illinois, and now owns a successful high voltage electrical equipment shop in The Woodlands, Texas. JR is a family man and works with his two sons and his wife. I admire that. You just don't hear about families working together anymore. JR once donated a kidney to his son who was badly in need. That's the kind of person he is.

JR called me many years later, 2010, I believe, to thank me for writing Fruitcake Hill and to tell me he had purchased copies for his entire family. But I had just emerged from a tooth extraction and was under anesthesia. Needless to say the quality of this exchange was compromised because gauze was packed in my mouth. I remember trying to say "I can't talk very well" and "They're trying to silence me."

I occasionally see JR in Texas. His home in The Woodlands is only a half hour away from our home in Spring, Texas. In a recent visit, I learned of the passing of JR's father to a stroke and his younger sister to cancer. We are all marching toward the light, so to speak. So treasure the memories but continue to make new memories. Life is short.

JX

JX was a Polish kid who attended grade school and high school with me. JX's mother had polio as a youth and it crippled her badly. JX's dad worked downtown at the Post Office and had immigrant written all over him, although I suspect he was 2nd generation in the U.S. He had a gruff exterior and smoked a pipe.

JX liked hanging at our house because it was never boring. We hung out in the basement of the new addition and listened to sports events on the radio. My cousin, Tarbender, called JX *Jetox* to mock how slowly he ran. And when JX tried out for quarterback in high school, the coaches had a field day with him. But what I remember most vividly was the day a car JX was driving accidentally struck a young girl walking with her friends down Roberts Road. Driving conditions were not great, I understand. And for months thereafter, a pall of silence fell about the incident. It was not OK to discuss this.

While JX was working to clear charges against him, he enrolled at the University of Wisconsin at Milwaukee to study engineering. Soon thereafter he dropped out of school, so he was around when I came home each year on Spring Break. One year, perhaps 1971 or 1972, JX and I went to an Adam and Eve contest at an Indiana nudist camp. I remember playing volleyball on the front line of our team and directly across from me was this gorgeous brunette, naked as a jaybird. The next year at Spring Break, JX drove his Corvette to

Florida. En-route we stopped at a Tennessee fireworks stand and purchased what we thought was a gross of M80's. Then we fired up his high powered Corvette and cranked up the stereo to the Who's hit song "Going mobile." And cruising down the highway we opened the driver side window half-way, depressed the cigarette lighter, lit the fuse and threw the bomb in the direction of the window. But the bomb hit the top of the window and bounced back beneath us. So there we were not knowing where the bomb was located and arching our behinds up to get away from the seat. Bracing ourselves for the blast which certainly would blow the windows out of his car we veered to the left and right. Shortly afterwards a powerful rush of dense pink colored smoke filled the vehicle. The M80s we purchased were M80 smoke bombs! We could see nothing at all through the windshield of the vehicle and luckily stopped safely on the shoulder. Once the smoke cleared, we could see the shocked faces of drivers passing by. And we had survived another experience by the grace of God.

On that same trip to Florida, JX took a wrong turn near Chattanooga, Tennessee and when I awoke we were 50 miles from Birmingham, Alabama. We pulled into a diner, made our orders and then witnessed the ugliness of the south at that time. The owner approached a young black man, perhaps 15 years old, and smacked him about saying "What did I tell you about coming in my restaurant?" The other patrons nodded their heads in approval. I turned to JX and said "Let's see this place in our rear view mirror!"

We made one additional stop at a Sea World in Florida and had a pretty good laugh. Sea World had a bottle nose dolphin swimming around an outdoor tank and we were told the mammal would surface for a Sea World-approved snack. But JX saw a different opportunity. He tapped on the water surface and when the animal rose, JX splashed it with water. At this point the dolphin banged its chin on the surface of the water and sent a reciprocal splash on JX. To JX, this was war. So JX again patted the surface, and the dolphin rose for his snack and JX splashed it again. The dolphin, at this point, sank below the water, turned over and its tail rose and came down hard near JX, sending a titanic wave on him. He asked for it.

The very next day we were sitting on the beach in Fort Lauderdale. Beach Boy tunes were being played everywhere and we were living and enjoying the freedom of Spring Break. But kids from Chicago, especially fair skinned Irish types, like me, were pasty white from the long winter and no match for the intense Florida sun. So by mid-afternoon I was a broiled lobster and suffering considerable discomfort. My ankles had swollen terribly and I was burned so badly behind my knees I could not stand erect. I tried golfing with JX the next day but it was unbearable. So the following day we set out for home where I could go to the doctor. Of course my sunburn had ruined JX's vacation and I felt terrible about that. And in the car on the return visit we did not talk much. He was mad and I was miserable.

I checked into the hospital on my return and was diagnosed with severe sunburn. The hospital administered several IVs for re-hydration and packed my raised legs and arms in topical burn crèmes. I spent three days in that hospital. JX didn't visit, as far as I remember, but my girlfriend from Western Illinois did. We ultimately married, and JX and I went on separate trajectories.

JX, at this point made the decision to move to Houston. I called him occasionally but it seemed as though he was now living on the moon.

Years later, I was hired as a geologist by AX and relocated to (you guessed it) Houston. This is where I next encountered JX. He and his new wife picked me up for an evening of food and dance. But something was amiss. And sure enough, JX did leave that woman and the next, and perhaps a third. Changes such as these you are likely to hear about in Hollywood.

I contacted JX in 2011 and stated I would like to pay him an amount commensurate to a RT ticket to Fort Lauderdale as reparation for the disastrous *sunburn trip*. And so I sent him 400 dollars to clear my mind. JX sent me a "Certificate of Absolution" stating essentially that my sins (to him) were forgiven. This is just what the Catholic in me needed to hear.

I still feel close to JX. The biggest impediment to us being even closer is his penchant to embellish stories. You never knew for sure the real story. I was overjoyed however, when he recently owed up to me that things weren't so rosy. I wish him well.

Discussing a character issue with someone is a bit like discussing how someone else snores. Often the snorer is not aware because he never hears it himself. He has to accept indirect evidence like others telling him he snores or waking with a dry mouth. Such was the case with JX.

JX's story does not end here. He also left the Catholic church and became a Mormon. Now this is no big deal to me, personally. I have many Mormon friends in Texas. But had his mother known, she likely would have turned over in her grave.

Wyn

Another character of mention was Wyn. My brother Lui called him Fort Wyn and Tarbender called him Walking Jesus. Wyn was a high school wrestler. I remember lifting weights with Wyn and was impressed how calm and good natured he was. I was always in perfect peace with Wyn. Words weren't necessary. He was easy to be with, loyal, and sported a beautiful head of hair.

Wyn had a gorgeous white Ford convertible with red leather seats. He would occasionally put the top down on nice summer evenings and cruise the vicinity of 95th Street and Cicero Avenue, sometimes picking up his friend after work at Browns Chicken. Wyn also owned a 3-wheel chopper motorcycle and looked the part of a biker dude.

Wyn and I and JR once visited a club in Frankfort, Illinois. Wyn agreed to drive his cream colored VW Beetle car but there was only limited space in the back seat and that space was reserved for me. And as Wyn slowed his car for a left hand turn into the parking lot, a car rammed us from behind. I was lying on the back seat and upon impact, the car deformed around me. It was not easy to get out but I did manage. And as it turned out, the kids who rammed us knew the sheriff and all charges against them were dropped. In the meanwhile Wyn's car was totaled and Wyn had no choice but to pay for all damages himself. Wyn was used to being stopped by cops because, as the police would say, he "fit the description." The paradox here was although Wyn looked the part of a trouble maker, he was the most law-abiding and kindest person I've ever met.

Wyn and I worked with his step father on weekends digging ditches and laying sewer pipe. Wyn also worked with me at the City of Palos Hills Public Works on the garbage trucks. At other times we hung out in the city lift stations and other places flying low to avoid detection. Once, several Public Works workers attended the children's matinee in a nearby town. Did we look out of place! Four big guys taking time off work, sitting in a sea of children! That would trip the alarms today, eh!

I am certain today that the real lousy jobs I had in my summers were my biggest incentives to go on in school. I remember distinctly that around August 10 each year when I resigned from these jobs I gave thanks I had an option. Bad summer jobs were critical in convincing me to stay in school. I didn't need more of that to know I didn't want to do that job forever.

Wyn did not go to college but took a job with Tinley Park Mental Facility. He always had the funniest stories of escapes and incidents in the mental facility. He also worked at a juvenile detention facility (jail) in Joliet, Illinois. He has always had interesting jobs and is the life of the party when we get together.

Today, you can usually find Wyn sitting in the Denny's restaurant on 95th Street and the Tri-State Expressway for breakfast. He is as constant as the sea.

JB

JB, of Czech descent, was raised in a family of ten children. JB was a year older than I was and graduated in the class of my sister, Julie, JR and Wyn.

Like Wyn and Shugs, he was a high school wrestler. He once broke his ribs in a match. Many in attendance heard them snap. But JB's coaches taped him up between sessions and he still went out and soundly defeated his opponent. He also practiced boxing with me and was a good sport in doing so.

JB was, and remains, a fine musician. He played his trumpet and trombone in a number of brass bands, his musical genre influenced by the bands *Chicago*, *Blood Sweat and Tears*, and *Santana*. He took a road tour at the height of the racial divide in America and remembers an evening when a black musician returned to the bus drunk, put a pistol in his face and asked the question "Why shouldn't I kill you, white boy?"

JB had a very serious heart attack several years ago and survived despite a series of cardiac crises in which the attending physician thought he had died. His recovery was a miracle. JB has been married for nearly 40 years. He has nine grandchildren and works for an ad agency in Will County, Illinois.

Pyts

Pyts is a nickname we gave to a friend I knew in high school and college. Some of the guys called him Olive and some called him Snake. He played against me in the church softball league and with me in our team known as the Piedmonts. He was a good fielder but did most of his damage hitting the ball. He was not as big or strong as I was but he had wrists that snapped off line drives. Pyts and I also played on a baseball team called the Maywood Stars.

Pyts loved it when we went out dancing because we resolved we were not going to let women's rejections ruin our night. We would dance by ourselves, in groups, or with other guy's dates. We were there to have fun. Pyts particularly liked it when I did my splits in the middle of the dance floor. I did another Elvis-like routine consisting of a stiff right leg, a tapping left leg, with a quiver of the lip. "Thank You, Thank You very much!" Maybe Pyts lived vicariously through me. I fed on his applause and he fed on my courage. To this day I can meet Pyts and say 'Hip Hip, It's all right!!" and this tickles him so.

Pyts kept in shape over the years and still wears the black leather vest I gave him some 40 years ago in Chicago. I could never fit in that vest today. So when I see Pyts it's like entering a time warp. He is still the same. And I like that about him. Like my friend Bri, Pyts liked Johnny Mathis and we would take dates to the concerts. We usually ended or began the evening with a meal at KonTiki Ports in Chicago, a truly memorable eatery off Michigan Avenue. We had other interesting times downtown, like the time I dressed in my Super Fly outfit of muck luck boots, purple see-through shirt, purple pants, black leather vest, Giovanni boots and dark shades. As I remember my friends walked a few steps behind me. No worries! I had plenty of courage, but perhaps I could have used a bit more discretion.

Pyts married a nice girl and together they had four children. I met Pyts several years later and shared a few beers. I noticed whenever I mentioned his dad he never followed with a compliment. Something was amiss so I asked him what the deal was and he said, "My dad was a contrarian. If I said white, he said black, when I said yum, he said yuk. He never gave me any credit and never learned to just listen." That's too bad.

JS

I met JS while playing baseball in Lemont, Illinois. JS loved hanging with us Palos boys and we became fast friends. He had the most unusual eyes, one green and the other brown and he loved talking about my junk automobile, the Lancer. He called it "The Glancer." It was a party on wheels and there were many times we drove the beater through the swamps just hell bent on having fun.

JS particularly loved a story about me meeting a girl's parents in Chicago. You see, the Lancer had no front passenger or rear seats because my father removed them to get higher gas mileage. As I remember the story went like this…

I parked the Lancer in front of my date's home and approached the door. And after a few introductions my date and I were off. It became apparent her parents were watching because I saw the front room curtains peel back. As we approached the car, I stepped aside and opened the passenger door. But to my date's surprise and her parents' horror there were no seats. In fact, all I had were blankets over the entire car floor. So I asked her to sit on the floor and closed the door behind her. And as I walked around the car to the driver's seat I couldn't help but glance at the parents faces. They were aghast! We boys loved the shock value of our actions and laughed about this for years. Poor Harriet!

JS went off to college at Southwest Kansas State and although we planned to visit each other in our respective college settings, that never happened. I had a girlfriend by this time and my life trajectory had changed. But I miss JS dearly and would love to contact him again.

The last thing I heard about JS he was fired from a Denny's Restaurant in Tucson, Arizona. Apparently he was the store manager and got into a fight with a drunken patron. I have not been able to locate JS although I think he may be living in Indiana or California.

Homer

Homer was my cousin. He was three years older than me and one year younger than my older brother John. Homer loved hanging at our house to hunt, play football, and occasionally go to movies. He was a movie aficionado, and could recite lines from a number of movies. I was usually around on Saturday evenings when Homer came to recruit his movie mate. The evening often ended with a stop for some pizza. All this kindness, of course, meant I was being recruited to help Homer sell papers at the church the following Sunday morning with temperatures expected to plummet below zero!

Homer commenced his high school studies at Sandberg HS in Orland Park but soon thereafter transferred to a catholic high school in Chicago by the name of De LaSalle. The school was located in a rather tough part of town and I remember him once wearing a large bandage from being hit over the head as he walked past some street ruffians near the school.

I was rather close to Homer. He and I spent a few years attending catechism classes on Wednesday evenings. One evening as we turned north on SW Highway in Orland Park, Homer, JR, Julie, and I nearly crashed head-on as Homer attempted to overtake a slow moving vehicle on a hill in a no pass zone. Luckily we spun out and avoided the crash. The Lord was apparently not finished with us.

Homer's two sisters and brother live in Arizona. The younger sister, an accomplished artist, learned in 2008 the fate of her cat

29

when she read Fruitcake Hill. Was she surprised!

Homer, his friend Mickey, and I remember the night of the Big Snow in 1967 when we decided to go tobogganing. We laid a toboggan atop the roof of Mickey's white Pontiac. I then climbed on top of the toboggan to hold it in place despite a biting arctic wind. You could imagine the sense of cold with the car travelling at some 25 plus miles per hour and an outside temperature of maybe zero. I would scream every couple hundred yards to get inside the car and warm myself. We never got there. It was too cold and there was already 26 inches of snow on the ground.

When Homer's mother died recently she left sufficient monies for Homer to purchase something he always wanted, a chopper motorcycle. Homer occasionally drives his motorcycle down our dusty road to visit Babe. Homer is a family celebrity.

Bob D

Bob D was another character that hung around our house. I believe he was older than my brother John but younger than my brother Frank, suggesting he had no direct connection with us. So I conclude he must have hung around to hit on my sisters Julie and Janet, both at that time eligible maidens.

Bob D, however, liked to hunt, had lots of guns, could fix our leaky pipes, and was always a warm body for a tackle football game. So he emerged on the approved list.

Bob wore glasses and had a bit of a pouch. He was also a smoker and sported brown stained teeth. Making matters worse he had one tooth that was broken. His appearance drew fire from another person named Small Paul and these two went at it regularly. They called each other every name in the book. One might start off saying "If I had a face like yours I would hide." And the other may respond "You don't have a face, you have a door mat." And the insults continued ad nauseum. But despite Bob's rough and unpolished appearance, he had an intellectual bent and taught my younger brother Billy the game of chess.

Bob bought a fancy car, a Torino I believe. The car was predominantly white with bold orange stripes running down both sides. It had high performance carburetion and other advantages. But because Bob was employed as a union plumber, his beautiful car became a mess in no time, storing his plumbing supplies.

I lost track of Bob when I went off to school in pursuit of my BS Degree. And Bob, at this time, entered a destructive relationship, marrying an angry woman who later shot him with a gun then poured concentrated plumbers drain acid down his throat and lungs while he lay dying on his couch. Bob passed away April 20, 1992 at age of 46 years. His wife went to prison.

Gerry's Professional Journey

I realized my calling to be a geologist in high school. I remember packing a thermos of hot chocolate and a few snacks after a fresh snowfall and walking to the top of what we called "Sullivan's Hill." There was a clearing atop that hill that provided a breathtaking view of the Sag Valley and its flanking ridges. These ridges exhibited some 60-80 feet relief and, according to scientists at the Illinois Geological Survey, once contained the waters of glacial Lake Chicago, a super-Lake Michigan. This meant, of course, the lake that native American Indians saw was different than the lake we see today. The ability to interpret the world fascinated me and I was on my way to becoming a geologist.

Perhaps my greatest influence for acquiring knowledge, however, came from a set of World Book encyclopedias my parents purchased when I was in elementary school. It seems like I read (or at least browsed) a book a day in summer months. And the glossy, gold edge pages of these books contributed greatly to my love of books.

UICC

I attended the University of Illinois at Chicago Circle (UICC) my first two years of college (1969-1970 and 1970-1971). I also played college football for the fledgling UICC *Chikas* program. To my knowledge, only a few of the players on the UICC team were scholarship players. Our competition consisted of schools like University of Wisconsin-Milwaukee, Ferris State, Wayne State, Delta State, Eastern Illinois, Illinois State, Kent State, and a few others. And we scrimmaged each year against two local schools, St. Procopius and Triton Community College. I once peeled off a run of 65 yards against Triton.

The following discussion paints one memorable picture of the times that frame that experience. In the fall of 1969 I enrolled at UICC and joined the football team as a walk-on kicker, wingback and fullback. We played our home games in Soldier Field, the same field that showcased so many Bears and Packers games. And I met a number of very special black people on the team in the following two years that changed my life. These guys were thrilled to know me and I was thrilled to know them. It was so red flag for me to actually be friends with these black ball players. This was the height of tensions in black-white relations and I was hanging with them. They called me their blue eyed soul brother, and I was quite proud of that. I actually introduced one of them to a nice black girl named Veronica that I met at the university Newman Center. And Tarbender, the giver of all nicknames, quickly began calling me Veronica.

Al

One of these black kids was named Al. He told me he was going to be a brain doctor. He furthermore confided to a black linebacker that he need not worry, because he would first need to locate a person's brain before he would operate. And he was convinced this fellow had no brain. So he was safe! Al's favorite saying was "You coconut head!"

Al lived on a seemingly dangerous street corner in Chicago, in the vicinity of Ashland and 111th and I worried about him. I went to his home on several occasions and his mother was so happy we were friends. Perhaps if I could stick it out with Al, perhaps he could make his way out of the streets and become the doctor he said he would. He and I dreamed of a better life and I am grateful he was there for me.

One evening, as we drank a bottle of cheap wine, we hatched a plan to meet each other and I invited him to come to my house in the country. My invitation was squashed when my father heard I invited him and said "don't bring a n….r to this house." That crushed me more than you can imagine.

I called Al that night and was too embarrassed to tell him the whole truth. I realized, however, that I must make an ugly compromise, if for no other reason than to insulate him from a more unsightly experience. I felt such shame. Al deserved much better. For weeks I had been the good ambassador for whites to the black community and when the real test came, my family failed me big time. I felt so exposed, so cheap, so hollow.

Al and I decided to meet the following morning at a diner in suburban Worth, Illinois. It was a half way compromise. We could meet and go our own ways afterward. It was so awkward. He knew, and I knew, there was a reason we met on neutral grounds that day. And he and I also knew that reasoning did not pass the smell test. That morning in the Worth diner was perhaps the lowest point in my life. And I pledged, at that point in my life, to choose another path. I would be different, I would be inclusive, and I would embrace my black friends regardless of the views of others.

Many years after the Worth Diner debacle I asked my father if I could swab the inside of his mouth to determine his family's matriarchal DNA. He responded positively, although with some trepidation and remarked "Don't look too deeply into this DNA stuff. You may discover something we are not proud of." I am convinced, from the words and the context of that statement that he wanted DNA testing to indicate his family did not originate from Africa. The DNA test, however, showed his Germanic/Czech people did originate in East Africa and migrated north into Yemen and west to Europe. Bob didn't want to believe it.

Bigotry was rooted in the country, our local community, and in our families. Looking back, Bob (my Dad) was a product of those times. Bigotry was in fashion and men talked in terms of those blacks, those communists, those Protestants, those Jews, and those Cubans. Someone had to take the blame for the country's social problems. And white flight from the cities to the suburbs was taking its toll on the city's vitality. In fact, my grandmother's bungalow in south suburban Beverly was sold on her passing for a mere 20,000 dollars, while the property today could easily sell for more than 300,000 dollars. And against that racial milieu was Al's and my friendship, struggling to be independent of the hostilities and distrust of the day.

I loved my Dad but he had one side that still needed some work. Bob had fantasies of escaping the racial problems in America by moving to Australia and took an extensive trip down under to size it all up. I don't remember what killed that dream, but he just stopped talking about it. He no doubt also noticed that Australia had its problems as well.

Miraculously, I contacted Al's son through Facebook in 2013. His son responded to my inquiry saying *"Yup, Al is my Pops"* and provided the email contact I needed. God works in mysterious ways! I have since contacted him and we have had lengthy discussions about the past and present. We both look forward to the time we can sit down and enjoy each other again. He calls me #47 and #31, my numbers in two seasons on the football team. I called him #21 for the same reason. Al was happy the UICC team was mercifully euthanized in 1972, since the away game at Delta State (Arkansas) was one in which the opposing team did not punt during the course of the game. You can't win like that! That performance was embarrassing!

In our lengthy discussions I thought it curious Al referred to me regularly as his white friend and I referred to him only as my friend. This confirmed his more difficult path as a black person in America in the 1970's and my more sheltered path in white America. That was a painful realization and opened my heart more fully to Al's life journey.

Dr. D

While attending the University of Illinois at Chicago, I met a great teacher, Dr. D. I met Dr. D at the very height of the Viet Nam student resistance movement (1969-1971). He was a new assistant professor in geography and I was impressed he had the courage to change his meteorology lectures in mid-stream to address more critical social matters, like the roots of political power and what students could do to resist the war. He even led a group from my class on a demonstration walk in Chicago. He sported a pony tail and was a maverick. I admired his courage.

I took one more class with Dr. D, this class focusing on astronomy. About two weeks hence, a well to-do classmate invited our class to a star party at his home in the rich western suburbs of Chicago. We were all outside looking at the stars through a telescope and smoking weed. Then it seems the party turned love-in and all I saw were butts bobbing in the flower beds. Dylan summarized it best "The times they are a changing."

Dr. D was a breath of fresh air for me in academia and his

methodology was affective. He changed one heart at a time and his methods focused not on material memorization but rather the lively art of argument. Dr. D did not believe grades were a measure of what you learned and instead allowed us to choose the grade we thought we deserved. Most chose A's but I took a B, because I didn't want him to get in trouble for giving all A's. In retrospect, that was stupid on my part.

I made a lot of friends over the years with my instructors. And why not?!! They were interesting people and I admired them. And they, in turn, also liked and admired me. I think this is because I knew they were people just like me and likewise needed friendships. With that most basic understanding, I nurtured friendships with former instructors well after my classes came to an end. I became their friends and by doing so, entered the rarefied atmosphere of academia. And it became more and more likely that I would exceed the expectations I placed on myself earlier in life. I could do it and I would do it. My teacher friends expected this of me.

Years later Dr. D invited me and my wife to his apartment on South Shore overlooking Lake Michigan. He and his wife had prepared a wonderful meal, the conversation was lively, and he was showing his collection of beautiful photographs he took in China. But that was the same night my beloved Nebraska Cornhusker football team played University of Miami for the national championship in college football (bad timing) and I repeatedly asked him to check the score. Luckily Dr. D admitted he attended the University of Nebraska for his Ph.D. and was likewise interested. Dr. D and his wife love photography and hiking mountain trails. Dr. D is still an active academician, leading cruises of the Chicago harbor and publishing on the same.

Brother John, Dl, and Bob F at UICC

My brother John was nearly killed in 1966 when a car blew a traffic signal on Cicero Avenue and 111th Street and ran him over on his Honda 165 cc motorcycle. John's helmet saved his life but he sustained a serious injury below his right knee where the vehicle's front bumper crushed his bones to powder. Babe was in deliberation that very evening with the attending surgeon encouraging the doctors not to amputate that leg. A year later he made a go at returning to college on crutches. Amazing!

My brother John and I took rides to the University of Illinois at Chicago Circle (UICC) with classmates Bob F and Dl. These friends had cars and we didn't. And both of these friends were typically late in getting started for morning classes that commenced some 20 miles away in bad traffic. So they drove recklessly down the Stevenson Expressway, sometimes on the shoulder, sometimes with abandon through traffic, and mostly a combination of the two. Bob's mantra was "Don't make eye contact, just drive." Dl piped in "Don't worry, the other guy has brakes." It's amazing we survived. Then there was the issue of parking once we arrived. It seems we were always forgetting the parking card!

Bob F and his live-in girlfriend once asked us to come for breakfast at his condo on Chicago's north side. So we agreed and left early in the morning for breakfast. On the way, however, we heard a news broadcast that a man named "feederitz" and his girlfriend were shot in an apparent armed robbery. It must be him, we thought. Later we learned Bob and Linda were shot a total of eight times and both survived. Apparently the .22 caliber ammunition the intruder used was old. Bob recalls he was taking the garbage out the rear door when he came face to face with a guy pointing a pistol at him. The intruder was in a low crouch and slowly stood up. He grabbed Bob's blonde girlfriend by the hair and went room to room collecting jewelry and cash. He then told them both to lie flat on the ground and he pumped four shells in Bob and four in his girlfriend.

Neighbors called the police. To this day I believe the shooter was encountered in a crouch position because he was looking through the keyhole, trying to get a look at Bob's sexy girlfriend. Bob surprised him by opening the door. The rest was totally botched and ended in a shooting. They are fortunate to be alive. Today Bob lives in Florida and lifts weights in hope of someday winning the award Mr. Senior America.

Bob F was my UICC boxing partner and although Bob was big and strong, things got ugly quickly because I was just much quicker. I also sparred with the center on the UICC football team. I had a talent for boxing and competed in the 1970 Chicago Golden Gloves Tournament as a light heavyweight (175 pounds). Dl was my trainer and we easily won our first match. Dl was four hours late picking me up for the semi-final match at St. Andrews Gym the next day however, and I ended up forfeiting. That was my biggest regret. I knew I could have smoked that fellow handily. A fighter knows.

Dl was a kid from nearby Hickory Hills and would do most anything for a laugh. One time, around Christmas, Dl and I decided it would be a good idea to get a Christmas tree. And armed only with a hatchet we surveyed a neighborhood until we spied our target, a beautifully shaped 6.5 foot blue spruce. I dropped Dl off and after about two minutes returned to the scene. All I could see as the car's lights fanned the scene was rapid arm motion whacking that tree. I made one more pass a few minutes later and there was Dl with tree in hand. But the funniest thing to us was not the tree but the stump left in the yard, and just what that owner would think when he woke up the next morning. Another time Dl stole a freshly baked homemade pie from a family style restaurant in southern Illinois. I will never forget him running to the car with that pie. Then there was the night the police picked us up in Hickory Hills. They asked us what we hit to leave a blue paint streak on the left side of his car and we explained we hit a red car on the right side. That was case closed for Hickory Hills and I am grateful to this day the probe did not go deeper. That boy was wild. And he just looked guilty. Today Dl has a large family and works in IT for a major university in Arizona. Dl is living proof that God works in mysterious ways.

Dad (Bob Sr.) and Lui on War Horrors

My brother Lui, a Marine, received a Purple Cross for sustaining an injury in Viet Nam and a Silver Rose award for fighting in agent orange-laced jungles. He spent several months in an Okinawa hospital recovering from these injuries and attendant psychological stresses. When he returned home, I remember him spending hours spit shining his dress shoes and his belt buckle but totally missing the point he should clean his room. Initially Bob Sr. (a WW 2 vet) and Lui (Viet Nam vet) argued about who had seen more war horrors, but later agreed they had different experiences.

Vietnam was hard on Lui and in time he became delusional. Once he told everyone he was going to buy a Mercedes, drive to Los Angeles and appear on Wheel of Fortune game show. He was picked up by police walking toward Los Angeles and returned to his home.

Today this war veteran lives in Wisconsin. He moved there because Veterans services and benefits were better there than in other states. He always liked the northlands but grew frustrated when he did not catch fish. In fact he used to say "Wisconsin, Land of 10,000 lakes and 8 fish." Lui just completed the fishing trip of a lifetime to Canada's Northwest Territories to fish for trophy northern pike. He has unfortunately been diagnosed with dementia and cancer.

Bob's Creative Energy Solutions

My father, Bob, was an exceptional man. I remember him as a tall man, perhaps 6 feet 3 inches with long arms and huge feet and hands. Those traits were passed to all his 8 boys. I liken Bob to a western movie hero and he even looked the part. And when he spoke, people listened (at least in courtesy if not in substance). Bob accepted his role as breadwinner for his 13 children and never complained the burden was too great. In fact, he found time to create solutions for society at large with creations like a bio-diesel Studebaker Lark, an automobile that ran on 4 or 8 cylinders for power-as-needed, a sod cutting machine complete with an auto-rolling device, a diesel powered home electricity generator, and numerous other inventions. He spent his own money and time trying to make a difference and attained the coveted mark of 450 miles per gallon of biodiesel fuel in a 10:1 mixture of waste grease to fuel oil, each gallon of the mixture pushing his experimental vehicles some 45 miles.

Bob's ingenuity was recognized and he won 1st Place in the 1980 Iowa Energy 500, a fair designed to encourage innovative energy solutions. Bob used to tease me that big oil should team up with him and "either buy him out or suffer the consequences."

Bob was deeply concerned about the pollution of earth's waters and atmosphere and in dwindling sources of hydrocarbons worldwide and believed the only solution was government rationing. The government, when Bob was a teen during WWII, provided coupons of two gallons of gasoline each week, and Bob in his blue collar patriotism believed the system was fair. I'm sure if rationing were imposed on us today, however, only the poor and the middle classes would be rationing.

Bob, Julie, and the Family Affair Band

My father, Bob, did not have good communication skills. His boys picked up on the communication deficiencies and always seemed a bit under duress in simple conversations. For example, they would wait nervously for times to interject non sequiturs like "Motley Crew" or "That's what she said last night" or "She's crazy about me." Strangely, I believe, communication problems contributed to his interest in forming a band. What he meant to say was he loved us and wanted to spend more time with us. The band helped him convey that message. Band members called it *The Family Affair*.

Bob learned to play the alto saxophone and clarinet as a child although he never took a formal lesson. In fact he never learned to read music either. He played by ear and was sort of a child prodigy. Bob's music allowed him to spend time with his own father and his father's drinking friends on weekends. His father had few kind words to say to him unless they were playing music. Bob's children learned the great classical works because music was so important to him. Bob cleaned up from the grease of the garage on Sunday afternoons and friends came over to have a few drinks and listen to the show. Bob loved making his own music.

Bob's big entertainment investments in those times were the purchases not of televisions but rather Magnavox home stereo systems. And he purchased complete record sets of the works of Montovani (his favorite band conductor), playing them unceasingly on his fancy units. We came to know The Nutcracker Suite, Clair de Lune, Eine Kleine Nachtmusik, and a number of Strauss waltzes including The Blue Danube. I can still envision Bob taking his precious vinyl albums from their sleeves and dropping the stylus needle oh so carefully. Bob found peace in this music. He had not, however, learned the joy of just sitting down to listen to his children. We excused him saying he was a product of the Depression.

Initially Bob and his daughter Tassy played together. She had formal training in music and earned a B.A. in trumpet and keyboard. She could read, write, transpose and compose music. This was good for Bob. Later Julie took a few months lessons on the drums and joined the band. Carolyn then joined playing tenor saxophone and clarinet and finally Jo Ellen joined on the guitar. And although Dad was the "Leader of the Band" and the best musician, the girls led the dialogue between scores. Babe, meanwhile, attended every engagement, usually with her sister Mame from Joliet. Together they encouraged and provided helpful criticisms.

The Family Affair played their first gigs for free at church picnics. Later they cast the marketing web wider to include weddings, corporate parties, bars, and other venues. Remuneration was not as important as exposure for the band. Bob just wanted to play and the girls were charged with securing the engagements. The boys were charged with setup, takedown, and transportation of the band. These were truly "family affairs" in that we were all involved when the band played.

Transporting the band required the purchase of a large GMC van. It was royal blue in color with a white streak running at door level front to back. Tassy drew a pink Panther logo on the back of the organ, matching evening dresses for the girls were made, and Bob dressed in a suit shirt, later upgrading to a tuxedo.

The band modeled their vocal score after the Andrews Sisters of the 1940's, the McGuire Sisters of the 1960s and Karen Carpenter of the late 60's and early 70's. These groups were excellent harmonizers and the four girls did quite well in producing similar polished musical pieces. They played a dinner and dance routine with music for seniors as well as young people. Polkas were their trademark. The band knew from experience the best way to get a crowd moving in Chicago is to play polkas.

The musical compendium of *The Family Affair* songs is provided below:

On The Sunny Side of the Street
On the Street Where You Live
Peanut Polka
Pennsylvania Polka
Chattanooga Choo Choo
Dream
Goodbye to Love
Close to You
A Song for You

Bist du Shon
Yellow Bird
Danny Boy
In the Mood
Proud Mary
Time in a Bottle
Copa Cabana
Does Anybody Really Know what Time it is
My Melody of Love
Blue Skirt Waltz
Celebrate
Big Bad Leroy Brown
Kentucky Babe
The Sound of Music
Has Anybody Seen My Girl
The Waltz You Saved for Me
Take me in your Arms, Dear
Snowbird
Saying Something Stupid
Boogie Woogie Bugle Boy
Killing Me Softly
Just My Style
Staying Alive
Got to Get You In to My Life
Melody in the Sky
Love Will Keep Us Together
Una Paloma Blanca
Let it Snow
Merry Christmas
Sweet Caroline
Stranger on the Shore

The Sunny Side of the Street, the lead-off tune, was a song from a Broadway musical that the Tommy Dorsey band rewrote with swing tempo to become a huge hit. This song became the trademark song of the *Family Affair* band.

The Family Affair played for approximately 10 years with Julie playing the first 5 of those years. Bob wanted to keep it going, saying the girls were on the cusp of stardom. They were, in fact, good friends with big band director Wayne King and the girls were, on occasion, invited onstage to sing *a cappella* between his numbers. But Julie wanted to marry and move on with her husband. Bob knew that losing Julie may signal the end of the band and threatened that he would not attend her wedding. But Julie resolutely stood her ground saying she was going to do it. Julie tried to repair the damage and asked him a few days before the wedding if he would walk her down the aisle. He said "No." And just as he said, Bob was a no-show for Julie's wedding. This was his loss. When the crowd returned to the house following the wedding, Bob was angry and on edge. He knew this looked bad for him, so we gave him some space.

At this time, Julie and her new husband considered themselves family outlaws. Siblings sided with either Julie or the band. It was a time of high anxiety. Bob didn't help his cause because he did not know how to negotiate. It was his way or the highway. Bob thought Julie was selling the band down the river and abandoning the family. "Marriage", Bob said, "was a fine institution but Julie should be able to manage the band, be a full time teacher and be married." But Julie considered the band a hobby while Bob and Babe wanted it to be a career. What started out fun, Julie said, turned sour in time.

Julie quit the band in February of 1976 and Bob had a massive heart attack the following December. Bob, like his mother and father, suffered from high blood pressure and this may have set off a chain of events that triggered the heart attack. Bob was under stress to be sure. And when Julie left the band, Bob and Babe repeatedly asked her "when are you coming home?" This was not an appropriate parental response.

This story is a classic example of losing face. Bob laid out his ultimatum and his daughter chose to ignore it. Babe, herself, took the side of her husband because "she had no choice." She said "I had to please him." I asked her many years later if she alienated her own daughter by taking that stance and she said "I suppose so." Babe, it must be said, was not the only one trying to please him. Heck the entire family sought his approval.

In retrospect it seems like protecting our father was the family's great mission. Bob may make a criticism to one of his children and Babe would rush into the room to smooth things over. The family then tried their best to pretend the incident never happened.

Julie's husband, sometimes blamed for Julie's departure, encouraged Julie to stop by her father's house and say hello when they travelled from his Coast Guard work in Cleveland, Ohio to Palos. His heart was good and so were his intentions. They endured a few barbs with every visit, however.

The band continued to perform five years after Julie's departure. They hired a young man named Lee and he became the drummer. Bob and Babe liked him and subtly suggested he and Kathy get together, but that did not happen. And to the end Bob and Babe tried to make the band a success, even interfering with the normal course of things. Of course they wanted a better life for their children than they had, but these things can't be engineered.

Basically the band broke up because it became too stressful for the family. More work was required of every performance than the parents and children were willing to expend. Julie was married and Tassy had a boyfriend. So the writing was on the wall.

The band and the human drama surrounding it are bittersweet memories for Julie. The one thing we all learned from this battle is that you don't give ultimatums to your children. You are left holding the bag if they cross the line.

Both sides had their justifications. On one side was the entertainment tenet "The show must go on." This is a powerful driving force and Babe and Bob certainly felt this pressure. And they must have been frustrated that Julie did not. The resolution of this conflict, however, should have been a simple value judgment, namely "Which of these do you value the most, i.e. Julie or the band." I am convinced my parents would have resolved this issue more quickly had they framed it in these terms. But when we are in the midst of battle the field gets smoky and your own people often get hurt. In the Army they call this "friendly fire." Ann Landers wrote an article in 1988 in an attempt to wake fathers from their destructive behaviors. Fathers, she said "just didn't get it" and repeatedly engaged in activities and actions that distanced themselves from their children and their wives. Dad demanded obedience but Julie stood her ground. Dad, of course, did not like the outcome, but I believe Dad respected Julie for her resolute conviction.

WIU

I transferred to Western Illinois University (WIU) in the Fall Semester, 1971. My mother, Babe, and her sister Mame drove me the 200 plus miles to school and left me there. I had no idea who my roommate was until a bearded, hyper young man approached me and said "Hi my name is Eric. I just escaped from the mental institution!" Apparently Eric was arrested on the streets of Miami giving his money away, and for that, his parents committed him. Eric loved being a bit off center, however. His major in school, after all, was Philosophy. He would sleep on an open text book, believing information would transfer to his brain without putting the effort forth to study. The last time I heard, Eric got married, moved to Tucson, and took a job as a mail delivery man.

Bill was my next roommate and we had so much fun together. Bill and I and our girlfriends once camped overnight at Lake Argyle. Bill got up early to fish but soon lost his lure on a sunken log. So Bill stripped his clothes, less his shorts, and swam to retrieve his precious lure. Bill was a dedicated fisherman. We also loved walking the fall forests for morel mushrooms and puffballs that we cooked in the dorm kitchen.

WIU was a perfect solution for me. WIU was dominated by farm country and the La Moine River. It was close to home, but not too close, and there was interesting geology close by in Missouri. But most importantly the bars served beer to minors. The availability of beer made the place just heaven, especially on Friday afternoons at the happy hours. Additionally, dorm floors organized bus trips to the lake for what we called "lakers at Lake Argyle State Park." I don't think laker parties are being run today. Too much liability, I suppose.

I decided not to play football at WIU largely because I realized I could not do a good job in academics and play ball at the same time. I earned a spot on the Dean's List for that wise decision.

Dr. Will and Dr. H

Dr. Will was a newly hired Sedimentologist at WIU when I arrived 1971-1972. He and his wife had an autistic child by the name of Billy. Neither Dr. Will nor his wife knew just how to deal with Billy. He would get violent, especially if he was not medicated. And in the end, Dr. Will and his wife committed Billy to a special school near Brenham, Texas, visiting him when they could.

Dr. Will and his wife took a particular liking to me and my girlfriend, Jean, and invited us on numerous occasions to dinner at their home. We became fast friends then, and remain so today. Will retired from teaching at WIU and now lives in the greater Houston, Texas area.

The students loved Dr. Will because he was so human. One time Will and I were on a field trip to Southern Illinois University and things were getting loud telling jokes and the like. And at this point a group of two or three professors from other schools came to our dorm room and asked us to quiet down. But Will, because he had a few brews responded "So what are you going to do about it?" Let's get ready to rumble, I figured. But that was the end of it.

Dr. H was another new hire at WIU, a petrologist with a Ph.D. from the University of Pennsylvania. He arrived about a year or two before Dr. Will. This man was quite the genius. If you asked him a question you may be obligated to listen to his answer for 10 or 15 minutes. Dr. H. took a liking to Jean and me. In fact we were the only students invited to a faculty party at his apartment where he served an eclectic cuisine that included chocolate grasshoppers.

Dr. H asked me and another student to accompany him on a trip to Colorado over a two week span in summer 1972. We didn't know at the time the real reason for this trip. We learned it focused on Dr. H getting away to decide whether he should marry a girl he was courting back in Macomb, IL.

On the trip, I attempted to shag some 20 pigs off a narrow one way street because our car could not pass the pig traffic jam. I also nearly killed Dr. H when I hammered on a large boulder about a hundred feet above him on a steep mine tip. That runaway boulder cut loose and passed him by inches.

Many years later, Dr. H advised us his son was suffering from a kidney disease and badly needed a transplant. A female cousin ultimately made the donation so all turned out well.

Both Dr. H and Dr. Will pushed and encouraged me to accomplish my goals in academia. It all seemed possible with my growing support and I began to believe, myself, that I could become a doctoral candidate someday and teach young collegiate minds.

Engagement and Marriage to Jean

In 1971 at Western Illinois University (WIU) I met a girl by the name of Jean. I asked her and a few of her friends to attend a party with me at a co-ed dorm known as Wetzel.

Jean's parents were from North Dakota and although they were no longer farming themselves, they were farmers at heart. My first encounter with the North Dakota locals took place soon after we were married in 1975. Jean's father drove a big 1973 Chrysler New Yorker from Chicago to North Dakota filled with myself and the family. The morning after our midnight arrival, locals had taken positions in the kitchen and the conversation went something like this: One of the relatives might say "Looks like rain" while the other, after a long pregnant pause would respond "Yep." And after another pause the first person may add "wind is picking up". To compound the situation, the locals talked in an accent a bit like the female deputy in the movie *Fargo*.

Jean's mother was a farm girl from Lisbon, North Dakota and had a number of famous euphemisms. i.e., Jees oh golly, the snow is deeper than ole Billy, Oh Nelly and others. She married a Danish farm boy from Lisbon who joined the Marines during the Korean conflict. Soon after he was released, he took advantage of the GI Bill and graduated with a BS in Engineering. He then joined Union Carbide, Linde division and managed a number of plant assignments along the margins of the Great Lakes. Jean's parents told us they chose a 2:00 PM wedding at the church because that allowed friends and relatives to go home and milk the cows. We're talking farmers here!

Jean and I dated for 3.5 years. During this time, I taught Earth Sciences in a Chicago suburb called Berwyn. Berwyn was the epicenter for Al Capone's mafia activities in Chicago in the 1920's and 30's. These suburbs were full of Sicilians and Italians (the family) and not so strangely, few blacks. The folks from these neighborhoods were not a very tolerant bunch. In fact the reason this job was offered to me in the first place was because a football coach at this HS shot and killed a student.

I soon realized I did not have the patience to deal with confrontation every day. Students with alarming frequency asked for the hall pass to go smoking, or whatever they were doing. So I pasted the hall pass on a 30 pound petrified log. Back then, this approach was funny and it did cut back on unnecessary trips to the restroom. A teacher may get sued using these same methods today.

Jean started a puppetry business when she was but 10 years old and continued to build her business, Marionette Playhouse, to this very day. In that time she never missed a performance. But the true mark of her effectiveness is whether the children remember her visits to the schools. And they do remember her shows. They tell her so in the streets! She has made a difference. And Jean has claimed every dollar she made as taxable income, even when the settlements were made in cash. I admire her honesty.

Jean and I were married in 1975. Surprises began immediately when our first night honeymoon suite at Holiday Inn was given to another couple. The good news was Holiday Inn management sent us down the road a few miles to another Holiday Inn and our Honeymoon stay was free. Later in New York City, we discovered our room in the Essex House Hotel overlooking Central Park was also given away. So the bellman went downstairs for a new key and they put us up on the top floor in the Presidential Suite, where President Ford stayed the previous week. The price was the same as the original room. So we were totally blessed on the honeymoon. And on the way home, Jean and I stopped in Herkimer, NY to hunt for Herkimer "diamonds." These crystals were not actually diamonds; rather they were doubly terminated quartz crystals that looked like diamonds. Today, Jean still has her crystals from that rock hunt in her "special things" box.

After teaching 3 years high school science in Berwyn I decided I should continue my studies with the long range objective of working in the oil fields of the Gulf Coast. I took Geology Summer Field Camp with WIU in the Black Hills and Jean and I spent our first summer together in a basement apartment in Rapid City, South Dakota. This was my first step towards an MS in Geology.

Jean, likewise, took graduate classes for her Master's degree in Family Counseling. One of her instructors by the name of Dr. Prof (I'm not joking) asked the students to take an index card being passed about and answer three questions. I don't remember the first two questions but I distinctly remember the third, which asked them to state the most unusual place they have ever had a sexual encounter. The students were then asked to pin these three answers on their shirt or blouse and walk around the room, displaying their answers. And of course they were stupefied when Jean's answer to the third question was "rock quarry."

I have always been attracted to brilliant people and Jean is a brilliant person. But she also has a heart of gold. She has never wavered to help others in need. She supported a sizeable loan to my parents to rebuild a rental property and provided financial support to two Indonesian families for many years. I am proud of her. We believe, with some reservations, that we are our brother's keeper.

Jean is also an amazingly talented organizer and served many years as the President of the Houston Puppetry Guild. In fact, when Jean was asked how she got interested in puppetry she always said it was because a puppetry troupe one day visited her elementary school. So I asked her what she would be doing today if she happened to be absent on that day, oh so many years ago and she responded "Then I'd be the President!" I believe her.

Jean works well with others as long as I'm not one of the "others." One amusing story comes to mind regarding her organizational skills. Jean volunteered to organize a neighborhood picnic at the end of summer back in Naperville, Illinois. She decided to invite an ice cream truck to stand by and sell ice cream to all those interested. So she stopped a truck, explained what she wanted and discussed the hours she wanted coverage. She stopped three ice cream trucks in all and told them the same thing. So on the day of the picnic we had three ice cream trucks standing by with three irate drivers, each of them thinking they would be the only truck on hand that day.

AX-Chicago/Ax-Houston

I interviewed at AX in Chicago in June, 1979 and fortunately was hired by Dr. Bill, who offered me the job then asked if I wanted to go with him to the health spa at lunch time. I thought that was a bit inappropriate so I advised him I did not bring my bathing shorts. Undeterred, Dr. Bill told me he never wears a bathing suit at the spa. Needless to say I did not attend. Dr. Bill did help me get my start in the oil biz, however, and for that I am truly grateful.

AX, at that point, sent me to Houston to train for a year. Initially I worked West Texas and later North Texas. There was much to learn about electric logs, mud logging, coring, drilling, offshore operations and calculating accurate water saturation numbers from logs alone. In addition, trainees were introduced to true stratigraphic thicknesses, true vertical thickness, reservoir engineering, seismic, mapping, and more. And despite the science available to us, it was amazing to find a number of reservoirs that were being drilled on statistical (i.e. non-science) bases, like the Sprayberry Formation, an incised valley trend. Managers would look at the performance of nearby wells and if there was any cause for optimism they gave their green light to drill. That was all the science there was on certain plays until seismic resolution was improved around 1982.

Well sitting for AX was largely used to pick core points and supervise logging operations. One time, the rig was in down time and being fixed so a number of drillers walked over to the mud log

52

trailer. It didn't take long for these deviants to decide they would test the volatility of their flatulence on a scale C1 (all gas) to C10 (all liquid). The drilling manager was first and he farted into the sophisticated gas identification tool and it tested all C1 (all gas). Next up was the roughneck and he tested C1 with a trace C2. And when the mud log engineer tested C2 and C3, the drilling manager said, "You should see the doctor, dude."

AX rookies spent a lot of time in course work in Tulsa at AX's training center. One of my colleagues was named Jerry. He and I became great friends in those early years. Jerry believed Houston to be corrupt money grubbing capitol of the oil business, and referred to Houston as "the pecuniary scene." Later Jerry had an opportunity to move to Denver and he quickly made that move. And to this day, Jerry has been in Denver, sometimes not in his best interests because when the oil price crashes, jobs in Denver were usually the first to go.

Not long after I returned to Chicago following my Houston training, AX announced they were leaving their Chicago headquarters and sending their exploration and production staff to Houston. We had just purchased a home in Chicago Heights perhaps two months previous and now had to sell it and find a new place in Houston.

The word from the top brass was if we could sell our Chicago Heights home quickly, we could move into a fine hotel in Chicago and qualify for expenses. This was one of my best memories. Jean and I stayed at the Palmer House and the Conrad Hilton over two months on expenses and met colleagues, relatives and friends downtown Chicago for many memorable dinners.

During the driving trip to Houston, however, we camped in state parks thinking I would impress management. Later, on submitting my expenses to the company, the accounting chief called and left a message he'd like to talk to me. I was nervous he would chastise me for spending too much. But he instead said "Look, this is a major oil company. You don't have to spend your nights in East Texas camp grounds." And I got the message. Next time it's the Hilton.

"House hunting" to my friend Jerry took on new meaning on his trip to Houston. Jerry said bagging a house in Houston required a lethal shot to the air conditioner. It seems like no one could survive in Houston if not for air conditioning, which appeared on the scene around 1932 but was not common in households until 1940 or so. Thank you, Mr. Carrier.

AX Trinidad

AX soon involved me in a structural and stratigraphic mapping project in Trinidad. I was really excited by the travel to the Caribbean, and the helicopter rides to offshore rigs. A Canadian colleague of mine once asked our supervisor for a trip to Trinidad and he said he would not authorize it. He would, however, invite him to his home to see his slides. This was not what my Canadian friend wanted to hear.

In retrospect, AX managers were executives who struggled to be geoscientists rather than the more usual case of geoscientists struggling to be managers. It was a suit and tie deal at AX, very formal.

In all, I worked three years on Trinidadian oil fields and visited Trinidad three times during the course of that work assignment. I loved Trinidad and Tobago's colorful pouie and poincianna trees, the scarlet ibis birds, the asphalt pitch lake, the island rums, the music, and the excitement of carnival.

But Jean and I realized I would need my MS degree if I was to make the most of opportunities in the energy sector and it became clear I should make the move back to college. So we planned an exciting Masters experience that would commence in St. Thomas of the US Virgin Islands.

St. Thomas, USVI (MS)

Preparations for my MS commenced when I made a phone call to a prof by the name of Charlie at Northeastern Illinois University. Charlie planned to spend his sabbatical in the US Virgin Islands. Charlie said I could help him in his marine studies and earn one year credit towards college classes. That was great, but where would we live? And where would Jean work? These and other questions remained unsolved at the time we arrived on this island paradise.

We arrived in St. Thomas in August, 1980. The terminal was an old military Quonset hut, open on two sides and cooled by fans. Outside, the temperatures were warm and the air humid. But on the good side, the palm trees waving to and fro made a comforting rustle. And small vintage DC-3 propeller-driven aircraft were parked everywhere, reminiscent of the final scene from the movie Casablanca.

Charlie picked us up at the airport in an ancient, small Rodeo-style pick-up truck. We sat on the front bench seat while the bags went to the back. He then drove us through downtown Charlotte Amalie and up the hill to his (and our) temporary home. The view from the top was just spectacular and we were awestruck.

We met Charlie's wife and kids and were shown the home and our room. All looked well at this point except there appeared to be some friction between Charlie and his wife over this whole arrangement. We agreed to pay them monthly rent.

A few weeks after we arrived, the gnats we called "noseeums" were getting intolerable so Charlie decided to pull the screen frames to the bay windows and take these in for repair. About a week later, Charlie and I picked up the repaired screens, put them in the back of the pick-up truck and headed for home. And no sooner did we leave the screen shop did we see three kids thumbing for a ride. So good natured Charlie, instead of remembering what we had just placed in the back of the truck says, "Sure, pile in!" And no sooner did he say that, he realized his error and watched in horror from his rear view mirror as his precious screens became unrecognizably bent. Oh did Charlie lose his cool! I still tease Charlie about this. Those kids, I am confident, had no idea why they got such a scolding.

Jean and I left Charlie's place within a month and moved to an apartment at Sapphire Village. That was the best thing that happened to us. Our view was unparalleled looking across the water towards St. John and Tortola and we witnessed the direct strike of a hurricane looking out these same windows. It was an exciting time. We did miss hanging with Charlie's kids, however. They really loved us.

Initially, Jean and I, being whites, were ignored in the local stores. But as time went on we smiled a lot and soon the checkout people waved us to the front. We won the locals over. St. Thomas was a Danish territory and had a slave history. Many of the men, we discovered, had 10 or more children with different women. The pattern was the men stayed with the woman until the child was no longer cute and then left them for another. In fact it was said the definition of pandemonium was Father's Day on St. Thomas!!

The cost of a liter of Crucian (St. Croix) rum back in those days was $0.95 while the cost of a liter of 7-up or Coke was $1.25. It was amazing that the mixer was more expensive than the booze. And so we drank up, in island fashion.

A curious phenomena in St. Thomas involved seafood. When we arrived, we were under the impression we could catch and eat all the fish we wanted. But it did not take long to learn that most of the fish sold for consumption was flown in from the US East Coast. Why, we wondered? This was because a kind of fish poisoning known as *ciguatera* was endemic in the islands and poorly understood. Eating a contaminated fish may result in death within minutes. The probability of death from ingestion of small affected fish was small, but the risk increased with larger, top of the line predators. So our diet, unlike our original plans, was considerably more diverse.

On the good side, I had at my disposal an 18 foot Boston Whaler boat with twin 35 horsepower engines and lots of lab equipment. And I dove (SCUBA) probably twice a week over a period of five months, the purpose of which was to map the damage to local reefs of an airstrip extension project. So sunburned was I from the diving that when Jean left her position as elementary school counselor at the end of the year, the children colored a picture in which Jean was colored black (in solidarity with the students) while my face was colored (you guessed it) red.

Diving with colleagues in the USVI

Charlie and I once put the boat in the water in a beautiful mangrove bay on the southeast side of the island. On the way out, we noticed two blonde girls sunbathing on the deck of on an anchored sailboat. They stood up to wave at us and had no clothes on, none! That was entertaining and startling, I assure you. About two weeks later I boarded an island jitney bus and sat next to a blonde girl. Just a moment later, this girl had an epiphany. She shrieked "Don't I know you?" I responded "No, I don't think so." "We saw each other while I was nude sunbathing" she cried. I said I was sorry I did not recognize her at first and advised the reason for the misidentification was quite understandable.

We earned half-price air tickets to any of the Caribbean islands from our association with the USVI airstrip extension study and used them to travel to many of the islands. One time, Jean and I visited Puerto Rico, a short flight from the Virgin Islands and I spotted a cave about 20 feet above a vertical highway road cut. Climbing to the top I encountered a number of beautiful stalactites and removed one. But how I was supposed to get down when I needed both my hands to negotiate the descent. So I asked Jean to catch this collector's item, and on the count of three I dropped the stalactite to Jean's waiting hands. She made the catch, but at a price because her hands got cut up from the sharp crystals on the exterior of the specimen. But I knew from that trial she had the passion for rocks that I shared and knew she and I would be a great pair.

While in St. Thomas we made friends with a really funny Jewish girl and her island boyfriend. She worked as the school nurse in the same school where Jean was the school counselor. She called the VI hospital a death factory and told us if we ever got sick to fly to Miami. Her boyfriend was a strong, good looking island native that we also came to love. He tried professional boxing and had a 17-4 record, as I remember. That impressed me so we boxed a few rounds. Jean and I framed this fighter's promotional picture and sent it to him in the mail. The tag on the back of the picture said "I can whoop this chump" and signed my name Gerry.

Upon our return to Chicago, Charlie invited Jean and I to a concert on July 4 in Grant Park. Charlie said he would be playing with the Chicago Symphony. We attended and discovered Charlie's participation was limited to firing the cannon for the 1812 Overture. Yes, it was a stretch, but Charlie did "play" with the Chicago Symphony.

Back to Chicago for MS

In June, 1981, after 11 months in St. Thomas, Jean and I moved back to Chicago permitting me to complete my Masters of Science (MS) studies under the tutelage of my old friend Charlie. We rented an apartment at 5800 N. Kimball and I commenced work on my thesis that focused on 80 cores in the fossil-rich Mazon Creek biota of the Francis Creek Shale. These cores were available at the Field Museum and at a core warehouse on the Chicago River. And after 1.5 years I wrote my thesis and defended it with no objections. Now that I had the MS, I needed to move on. I thank Charlie for being my advocate and advisor. He would always think in possibilities and I am grateful for his risk-taking on my behalf.

MO Egypt

We moved back to Houston and I accepted a job with MO, specifically working the Nile Delta and the Gulf of Suez, Egypt, and attended partner meetings in Paris and Cairo virtually every month. These were very busy times. In Egypt I learned *baksheesh* was the Arabic word for tip or bribe. And although the company insisted they did not give bribes, they did admit to giving facilitating payments. Hey call it what you like, but these payments moved our papers from this desk to that desk. So how is that not a bribe?

Our boss at the time loaded us with work. He used to say "Don't ask the mules, just load the wagons." I clearly remember one time this man met with us and said he would need a drilling montage and he would need it tomorrow morning, so get after it. Now preparing a drilling montage when none of the components were ready for inclusion is not sound practice. So I worked on my geologic part, another fellow worked on the geophysical part and various drafts people worked through the night making it beautiful. And by 8:00 AM, with final coloring completed, I proudly went to this man's office, showed him the montage and he said "Oh, I decided not to use that." Were we angry? You bet! He could have called off the dogs and let us all go home hours ago, but no, he worked us through the night.

In one trip to Cairo, a colleague of mine and I were sitting in the Charles de Gaulle Airport in Paris when an announcer asked if Gerald Kuecher would please come to the information desk. So my friend and I made our way to the desk. Upon arrival the attendant asked if I was Gerald Kuecher and I said yes and he presented me with first class tickets. My friend, awestruck by these proceedings asked, "Well what about me? I'm flying with Gerald." To which the attendant looked at his ticket and said "I'm sorry Mr. M, You are in a different class." And every time we met thereafter, I would say "But Mr. M, you are in a different class."

One geologist I worked with in Egyptian exploration was the quintessential know-it-all. I had just entered the Ramses Hilton Hotel in Cairo and there at the entrance was a huge statue of Ramses. I wanted to check it out so I went to the back and taking a key, made powder of the material used to make the statue. This of course meant the statue was made of very soft material, like gypsum, and was not a granitic antiquity. So when this fellow approached and commenced bragging the statue was granite and that he himself watched this statue dug from the ground, I showed him proof that was not the case.

My statement on business travelling follows: Yes it was fun to get to our destination, yes it was fun being on expenses, and yes it was nice to accumulate frequent flyer miles, but it was marginal to dreadful working all night to prepare presentations, going through security, updating our visas, waiting forever for delayed flights, hanging with the same people all day and all night, dealing with jet lag, taking required immunizations, being away from your families, and contracting various gastro-intestinal problems. By the way, the key to staying healthy in Egypt was to never drink un-bottled water or drinks with ice, and avoid salads and fruits peeled by hand. Eat only cooked meals. And still you may get sick, even if you eat at the best hotels.

I was asked by the VP of another region if I would sit a very important well off Kinsale Head, Ireland. I agreed, flew off to Cork, Ireland and commenced my operations i.e. two weeks on, two weeks off, two weeks on. In all, I spent 7 weeks in Ireland and got to know the isle rather well. I love the humor of the Irish, an example of which is contained in the following: "We're a small drinking community with a fishing problem." I would love to have a cabin in Kenmare, Ireland. I feel I can really get away in Ireland.

Jean and our new baby daughter came with me to Ireland. I remember leaving our daughter at the home of our Irish friends and walking across the street to the neighborhood pub. We asked "where is the babysitter?" and they said "Oh we do this all the time. Don't worry, the children will be safe."

When we walked in to the pub everyone came over to see and welcome us. The art of conversation is alive and well in Ireland. No TVs, no music, just drinking and conversation. I kept hearing "Hey Yank! Let me buy you a beer!" from all parts of the bar. I felt a deep communion with these folks. And to this day we celebrate St. Patrick's Day with corned beef, Irish soda bread (from cousin Mellen) and beer. It is a lovely tradition. And I frequently think back to the friends I made in that pub.

MO was the best company I worked with through 1985. What happened after 1985 is up for debate. I know that scrutiny on drillable exploration prospects suffered. One of the big bosses used to say "You can't find oil unless you drill." And although there is some truth in that statement, there also was a corruption of thoroughness. The message from above was the drill bit will determine if the petroleum system is complete, not the other way around. So we drilled "prospects" that were not worthy of the name. In addition, geologists and geophysicists often developed plays independently. Case in point a geophysicist once asked me why the exploration drilling results on his prospect was 1000 feet off in his interpretation but within 100 feet on my interpretation and I said "you never asked me." Additionally, and at the request of management, geologists ignored certain field observations, and mapped east dip in the southern Gulf of Suez where the dip was quite clearly west in all field observations, satellite maps, and dip meters. We did this to mature leads.

After three and a half years, MO elected to move me to their Denver Research Center. There was one snag, however, as MO's policy at the time required zero net gain, that is, I would go there and another person would take my place in Houston. Well I did not know this at the time but the transfer candidate and his wife were in the process of divorcing and they were not going to move at this time. So after considerable delay, that arrangement fell through. So now what? And by the end of the day, my bosses called me to a meeting and asked "OK, how about Jakarta?"

This impending move saved me. I was out the door in March, 1986 and moving to Jakarta when MO went through what was its most extensive layoff. And on July 11, 1986, the Houston headquarters laid off 112 workers and lost another 37 to early retirement. Company insiders called it Black Friday. In all, 30% of the technical staff was made redundant. But it was the way these people were let go that hurt MO's excellent (at that point) image as a family company. The morning the axe fell, dozens of taxis were parked along the streets to take the unlucky staff home. In addition, security guards escorted the affected to their rooms and watched them pack. This was a disgrace. If an employee wanted to steal something, they had plenty of opportunities before the layoff. I was glad I was living elsewhere when that mess came down or I surely would have been released. Corporate loyalty died that day. And employees realized MO embraced *dog eat dog*, just like every other oil company.

Life was good while MO had their research center in Littleton, Colorado. But that ended too when times got tough in the oil biz.

MO Jakarta

Jakarta is far from Houston. In fact, Jakarta is about as far away from Houston as one could get on planet Earth. We were 12 time zones away during one time of the year and 13 time zones away during daylight savings. Jakarta is also 6 degrees south in the Southern Hemisphere. Now if one needs proof of the International Date Line, consider flying from Hong Kong on Sunday at noon, flying through the night and waking for breakfast 12 hours later on Sunday morning in Los Angeles. If we all would agree this is really Monday, perhaps we would not be so confused. Flying time alone Houston-LA-Hong Kong-Singapore-Jakarta was 24 hours, and with stopovers this required two travel days.

My family and I arrived in Jakarta in March, 1986 after a brief stay in Hawaii. At the time we had one child and Jean was 2 months pregnant with another.

A MO company representative met us at the airport and took us to our temporary quarters. The following morning we were up and getting dressed for church when I caught a glance in the bathroom mirror of something large, dark and slithery crossing the surface of the pool. And sure enough, when I investigated, I found a 3 foot monitor lizard clinging to the drain at the bottom of the pool. I grabbed the skimmer pole thinking I could throw him out if I could plaster him against the netting by force of lifting. I tried twice and failed. But on the third attempt, the net broke the surface and I readied to pitch the beast over the wall when the critter jumped out of the net and directly on the back of my wife Jean. The lizard stuck the way only lizards can, and Jean screamed. The lizard then jumped off. I then cornered him and pitched him over the wall using the skimmer like any self-respecting superhero. Jean, was concerned, and for good reason.

Our Kids

Sanny, our oldest, was born in 1983 at a time I was on the move, professionally. She learned to crawl in Denmark and learned to walk while we were on assignment in Cork, Ireland. By the time she was 3 years old we were living in Indonesia and traveling throughout SE Asia. One fine morning, Jean and I and Sanny were sitting in the Hyatt Hotel restaurant in Hong Kong eating breakfast and Sanny was coughing every time she ate a spoon of food. I told her she should not cough after each bite and she replied "When we get to the ferry (our plans for later that day) you sit in the front and I'll sit in the back and choke." This same daughter, as a teenager, rafted the Grand Canyon with me and some 15 other geologists. And upon arriving at the site geologists call the Great Unconformity, (an unconformity is a place where significant erosion has occurred and the record in the rocks has been removed), professionals on board took the opportunity to share the importance of this site and we spent a large part of the morning there. When we were about to ship out, someone asked Sanny what she thought of all this and she said "Geologists are the only people I know who could spend two hours talking about something that was not even there!"

Larue was born in Singapore in 1986. Jean was flown to Singapore about 4 weeks prior to her anticipated birthing date because we wanted to have the baby born in the better quality Singaporean hospital. Larue, it turns out, was the only blonde in a nursery of black haired Singaporean babies so it was quite clear which baby was ours. I used to say *"Your hair is as perty (pretty) as corn silk."* Speaking of blonde hair, Indonesians had a nasty habit of pinching blonde children, the idea being they would have good luck if they did this. Larue, of course, did not like to be pinched so we did our best to shield her as we walked in crowds. In high school, Larue raised her own pig, Olga, as a 4H activity. This pig once bit the ear off another pig that ate her food. Olga was a pugnacious pig! Larue also had her own horse, a thoroughbred named Andy.

Dance and her brother Dane came to live with us in Indonesia when their parents (Jean's brother and sister-in-law) died in a terrible traffic accident in Missouri in 1988. Only two months before, Jean's brother and sister-in-law asked us to raise their two young children if anything grave should ever happen to them, and two months later, the parents did perish. Remarkably, both children survived. They lost their parents, their home, and all things familiar and moved to a place that was totally foreign and our family instantly grew from two children to four. That was a great culture shock for us all. Today these children both have easy going countenance and are a joy to us.

The accident occurred some 25 years ago and the kids are now grown. Our oldest girl is now a lawyer in Chicago, our second girl is a fashion designer in NYC, our third girl works as a purchasing manager in Houston, and our boy is studying to be a Physician Assistant. All the kids have found their way and we are grateful.

We were expatriates (expats) in Jakarta. And what is an expat? An expat is someone who is willing to live in a part of the world outside his or her comfort zone for the purpose of seeing remarkable things and making remarkable money. Expats exude a confidence they could survive anywhere, and could negotiate any situation in a foreign country

There were both good and bad sides of being an expat. On the good side, we had servants and more vacation days than folks back home. On the bad side we were semi-permanently isolated from our friends and families. But since all the expats were in the same boat, making good friends was seldom an issue. Everybody was lonely. And there was something about that arrangement that nurtured maverick personalities.

We had six servants (pembantu) in Indonesia. Having servants made our lives easy. We never had theft and had the best relations with the community, relations borne on trust. We had a yard boy who cut the grass with a scissors while sitting on a half-coconut, a guard (jaga), a cleaning lady, a baby sitter, a cook and a driver. The driver was essential because if you had an accident in this country and killed someone, locals could surround your vehicle and give you big problems, an eye for an eye.

Servants were great to have. I remember telling my cook on any given day we would be having a party that evening and she asked just two questions: how many people will attend and what do I want to eat. And after each meal was finished we could walk away from the mess and it would be cleaned spotlessly. Never again will we have it so well. Of course we were vigilant. Our leafy vegetables were washed thoroughly in bleach and permanganate; we drank bottled water, and still occasionally came down with gastro problems and boils.

The importance of maids and drivers to expat mothers cannot be overstated. Young mothers have much to do and much of it involves running errands. The servants represent a wealth of information on where things can be found in the city and help in communicating for the items in question. So it is not unusual that mothers running a household feel so overcome yet indebted to their maids. Some mothers, in fact, would say they owe their very sanity to their helpers. Nothing is simple in Indonesia. One must go to many stores to complete your errands as there were few one-stop supermarkets. At least that was the case when we were residents.

Servants had their own particular quirks, inherited behaviors from pre-Islamic days. One of these involved picture frames hanging on the walls and spirits. They believed evil spirits sat on picture frames and the only way to get them off, these nasty gremlins, was to tilt the frames. So we may go through a room and straighten frames, then a short time later we'd notice the frames were tilted again!

It is amazing there aren't more casualties eating in Third World Indonesia. It appears there is filth everywhere. But the people have learned to deal with rats, roundworm, mosquitoes, and typhoid. They boil their water, wash their hands, and take monthly medicine for worms known as obat cacing *(pr. obaht cha ching)*. In addition they have learned certain fruits are safer to consume than others. The banana, for example, can be peeled and eaten without ever touching the fruit. Other delicious Indonesian fruits include the rambutan, a hairy looking reddish fruit, the purple mangosteen, and the durian which is said to taste like heaven and smell like hell. Even dry products like flour cannot be trusted, and you may see the surface moving upon opening a sack of flour or pancake mix. The critters are usually weevils. Seeing this usually cures your fancy for bread, cake, or any other bread products. Jean dealt with weevils by placing flour and cereals immediately in the freezer upon returning from the supermarket. Then the dry products were sifted to remove bug bodies, if present.

Indonesia is a mystical place. The Indonesians were animists long before they were Muslims or Christians and today, they still maintain these earlier gods and beliefs as well as those of Islamic and Christian faiths. Evidence can be seen in the wayang kulit (leather shadow puppet) shows that play all night long with ancient Hindu god puppets while kuda lumping shows (horse magic) were quite dark and evil and capable of sending locals into trances.

Indonesians believed they could overcome sickness by drinking magic solutions. The lady that carried these potions was called the ibu jamu, which literally means *mother of magic potions*. She carried bottles of various colored liquids for curing ailments as well as coins for rubbing the throat in case of sore throats. An Indonesian recently remarked "These really work well in Indonesia." The corollary, of course, is they do not work well in areas outside Indonesia.

My driver surprised me one day by asking me if he can take off a few days to go 5 hours away and get cured for his ailment. "What ailment?" I asked. And after considerable prodding he told me he had an infection in his penis that he contracted from peeing on a sacred tree and pleaded with me that he must go on this journey to get it fixed. Of course we said, "Hey just go to our doctor" but he just had to go to the witch doctor. We let him go and he returned saying he was cured. This is how it was in Indonesia. The shingle hanging outside the local dentist office, in fact, had a picture of a pliers and a tooth; not sophisticated but effective.

Indonesian servants occasionally asked for money from us for paying their debts. My driver, in fact, asked "Tuan (master) I'd like to borrow 6 million rupiah." I responded "Just how do you intend to repay this loan?" I laid out a schedule that if he repaid 60,000 rupiah per month it would take 100 months to repay in full. Then I asked "And what would you live on in those 100 months?" He said "I'll borrow that from you." We did not lend him the money that time.

Indonesians have a very long history of borrowing. When spices were discovered in the Moluccas Islands, Indonesian locals did not want to work. So the Dutch opened a commissary and told the locals "Take what you want." And they did. And as they walked out the door they were presented with a ticket stating the amounts for which they happily signed, and sure enough they agreed variously to one and even two years hard labor.

The currency used in Indonesia was the rupiah. Because inflation was so high, it took a lot of rupiahs to purchase anything large. The largest bill used on the street at that time was a 10,000 rupiah note, equivalent to about 6 US dollars. Today, with hyper-inflation, that same 10,000 rupiah note is valued at just 1 USD. To purchase our SUV automobile, we had to go to the bank and get a large shopping bag full of money. Needless to say, we felt very insecure walking out of the bank and into the streets with all that money. The locals called us expats the orang kaya, meaning "the rich people."

The Indonesian people are very good natured and concerned

largely with making a living. They have a great sense of humor and enjoy a good laugh. We talked with them in the streets whenever possible. We, after all, believed we were ambassadors of the United States and never felt threatened. Indonesia is largely Muslim but unlike the Saudis, Indonesians are quite tolerant of other religions and it is not uncommon to have a church (gereja) located close by the mosque (masjid). In fact, during the recent Arab Spring uprising in Egypt, Coptic Christians from one church surrounded a nearby mosque to protect it from retaliation by their own. This is a special relationship, a model of how to get along in a complex world.

Ramadan is the Muslim holiday during which all faithful, able bodied Muslims are required to fast. The people fast between sunup and sundown and since those times vary day to day, many businesses provide alarms that signal the time the fast would end. And because the people were starving all day, they will eat at the strangest locations. One of our most interesting encounters with the Indonesian Muslims at Ramadan took place at the movie theatres. Once the signal was given that the Ramadan fast had ended, patrons opened pop tops and broke out the feast in total darkness.

Indonesia honors its women. Indonesian women have their own businesses and one woman was once President of the country. Woman's images also grace the national currency. Tolerance is modeled in Indonesia.

My western assumptions were tested on my first week in the office. I was advising an Indonesian geotechnical aide that I would need him to finish edits on a map by the following morning and he responded *Insya Allah* which literally means *God willing.* But to western ears he was saying *maybe I will and maybe I won't.* This was unacceptable, of course. I tried the same approach again and got the same answer. But as time went on I realized that *Insya Allah* is how Indonesia and the rest of the Muslim world operates. Maybe the photocopy machines will be inoperable and maybe they won't. You can't be assured of anything in Indonesia. They were right.

Jakarta is in a low basin and is hit by occasional floods. A letter from our driver, Amsar, tells the story of the latest flood in 2013.

Dear Mr & Mrs.Kuecher,

Thank you very much for your financial aid. Amsar received it and used the money.

You must have read or heard the news, that Jakarta was flooded on January 17th. Almost half of Jakarta was flooded, including Amsar's house. But thanks to Almighty God, Amsar and family were fine Amsar's children and grandson were fine when the floods hit the house.

The water level in Amsar's house is 50 cm, but in places such as West Jakarta, East Jakarta, North Jakarta, Karawang and Serang/ Tangerang, the water level reached between 100 and 200 cm.

The flood in Amsar's house was only one night, but the flood elsewhere may last a week. Amsar pray to Almighty God, may He give His blessings to them.

Floods that hit Jakarta this year were the worst, because a lot of heavy rain fell and paralyzed the land because it is, in places, below sea level. Many people who did not flood couldn't go for work because roads in Jakarta were flooded.

The rainfall stopped on January 18th, but the water level in Jakarta was still rising. A number of dams and dikes collapsed due to heavy flooding. Thanks to Almighty God, Jakarta is getting back to normal. No floods anymore, but still having traffic everywhere.

Amsar is really thankful for your attention, support and kindness. Amsar always prays to Almighty God.

May HE always bless & protect Mr, Mrs.Kuecher and family. Say my warm regards to your children. Amsar hopes they are doing fine and always are protected and blessed by Almighty God.

> *Warm wishes,*
> *Amsar and Lia*

The office environment in Jakarta was not good. This may have been due to the mix of characters. When MO's original exploration team broke up a few years later, a briefcase was found at the bottom of a cliff. And the word got out that maybe there was a suicide. The owner of the briefcase was found to be a geophysicist on our team. The owner insisted he flung the suitcase off the cliff because he was frustrated with the company. This was not a surprise to any of us. The company frustrated us all.

One time, MO participated in a special well drilled in the Malacca Straits. Upon logging, we discovered three zones that looked like water but had accompanying gas and oil shows. These suspect zones were found to be shaley and thin bedded. The operator did not want to test because these zones calculated water. But we (MO) persisted and the operator reluctantly agreed to test one of our three suspect zones. And sure enough, the thin bedded zone we proposed for testing produced 955 barrels of oil per day. Why? Because small amounts of clay suppressed true resistivity measurements some 50-60%. This was our first taste of shale as a reservoir and was a lesson well-learned. So, to address the question "What are we supposed to do with the Indonesian water saturation equation if we find hydrocarbons in shaley rocks only 0.5 ohms higher than background shale?" And my answer was "Test it." If the water saturation equation is proven inadequate to deal with these rocks then we need a better equation. And thereafter we tested every zone with gas and oil shows.

Craw

In Jakarta, I had a work colleague by the nickname of Craw who flipped his bean bag incessantly to "embrace boredom." When one would ask Craw what he was working on, he would say he had not worked for months. But MO would not remove him because it cost so much to bring him to Indonesia. So Craw ignored the everyday office activities and invested heavily in the stock market. The year was 1987 and stocks were rising quickly. Craw watched his accounts every day waiting for the "highest high." Then one day Craw said, today is the day and sent a sell order. But back in 1987, it took three days advance for a sell order to be executed from Jakarta. So Craw actually sold three days hence on what became the "lowest low." Craw lost a fortune over the weekend. But there was a simple elegance to Craw. He knew life should be more fun. Management would try to get even with Craw by saying his performance review was less than stellar, and Craw would respond, "Hey, I just don't care" and they had virtually no power over him. He had that "take this job and shove it" mentality. The company did get even with Craw, however, when he returned to the US. He was asked to come in to work on a Saturday and reportedly said "Ask someone who cares." Craw was fired on the spot. Today, Craw sells jewelry at the major geologic and geophysical conventions. Craw was one of the most interesting people I ever met.

Bingo Wally

We had a male dog in Indonesia and the kids called him Bingo Wally. Why they called him Bingo, I don't know. What I do know is this dog was hyper-active and would run as fast as it could around the perimeter of the pool and knock our children into the water. I would come around the corner and see my children flailing to stay afloat and glance over and see the dog smiling in his accomplishment. So I posted an advertisement at my office and within an hour's time a Chinese Indonesian came to my office and asked, to my shock and amazement "What does the dog taste like?" No, this cannot be. Certainly he must mean "What does the dog like to eat?" So I convinced myself he must mean the latter and made the transfer. About three days later, my wife and children asked me to contact the man so they could visit Bingo Wally. When I asked the man if we could come over to see the dog, he put his head down and said "I'm sorry, the dog ran away last night." Yeah sure! I'll bet. Chalk that one up to naiveté, I guess. I assure you I did not know the man's intent when we made the exchange.

Softball in Jakarta

Softball was king in Jakarta. We played two seasons per year and often had two games per week. We also had post-season tournaments that were an even bigger draw. Men, women, and children showed up to socialize. I was immediately recruited to join and captain the MO softball team that played in the B league. We dominated that league for about two years. Then our team was asked to join the more competitive A league. I was chosen for A league all-stars in 1988 and our team won the SE Asian tournament in Jakarta over teams from Indonesia, Singapore, Hong Kong, and Sumatra. It was a wonderful time.

Java Lavas

In Indonesia, I joined a group that climbed volcanoes. The group was appropriately called the Java Lavas. The folks in this group loved adventure and you can't find more adventure than climbing volcanoes. There were a great number of 13,000-14,000 foot volcanoes in Indonesia. Many of the tallest peaks required an overnight stay on the mountain. Smaller peaks were doable in one long climb. Arriving at daybreak provided the best chance to look into the crater.

Gunung Kawah Ijen located in west Java is an amazing place. The water in this crater lake has the acidity of strong battery acid and is near boiling point in temperature. Sulfur stalactites are mined daily by unknowing, unprotected Indonesian laborers and it may represent the world's most unsafe workplace. A volcanologist from Alsace-Lorraine (France), in fact, once drew intense criticism when he and another scientist inflated a rubber zodiac boat and paddled out into this yellowish-green poisonous environment. A magnesium metal chain affixed to a probe was filmed dissolving in a few seconds. Had the boat developed a hole, these volcanologists would have been consumed in a very short time.

Author atop Gunung Gede, Java,
with two other volcanoes in the background.

It's hard to say which of the volcanoes I climbed over the years was most impressive. But I suggest Kelimutu in Flores deserves mention. Kelimutu has five crater lakes, each at a different pool elevation and each lake a different color. Locals think the different colors reflected the moods of the gods. Any way you look at it, Kelimutu is totally amazing. So amazing, in fact, that when I approached it to photograph from the edge, I realized the place I was standing just moments before was nothing but an under-cut ledge and I could have been down the cliff and into the lake some 1000 to 1500 feet below. I think the government should place barriers back from the edge to save lives, but that's unlikely.

Author atop Gunung Kawah Ijen in Java with local security guard.
Note the gassy fumarole near the water line.

I was victim of a crime only once and this crime occurred not on Indonesian soil, rather in Malaysia. I had just finished teaching a Petroleum Geology course in Kuala Lumpur and took a cab to the railway station. A lady cab driver picked me up near the twin PETRONAS Towers and when we arrived at the station, she popped open the trunk of the cab, at which point men outside began taking the bags. The lady driver kept me inside making change and by the time I stepped out of the car, my bags, passport, tickets, and money were gone. This was a set-up. It took me three days to get out of KL with no papers. I learned a very important lesson here.

Geologists are an Odd Lot

Geologists are strange. They will go to the ends of the earth to collect minerals, fossils, rocks, rock structures, photographs, and rock samples. My good professor friend Gordy once told me a story of a heavy-set American geologist who travelled alone to Timor, Indonesia in the midst of a civil war to collect rare fossils of a very large species of *Pentremites*. The story began with the geologist boarding the train headed in the general direction of the fossil locality. And an hour into the ride, the train suddenly slowed and armed militia wearing ski masks boarded the train. Once onboard, bandits began shaking everyone down for money and jewelry. The over-weight geologist looked straight ahead but it was easy to see by his profuse sweating that he was under severe duress. One of the foot soldiers noticed the man and asked his boss (the general) to speak with him. The general knew English and asked the man if he was American. The man said yes and shared he was traveling to Timor to visit the site of the giant *Pentremites*.

The General then said "If you are looking for the giant Pentremites I will take you there. The giant *Pentremites* are near my house and I will be happy to take you." Doesn't this sound a whole lot like *Romancing the Stone*? So here we had a lonely scientist that somehow found another person who shared the passion for the rocks and was the leader of the bandits. Curiously, the geologist was the only person not robbed of money. Had he not found this particular educated person, the outcome would have been uncertain at best. Now what are the chances of this happening, one in a billion? The strangest things happen in the lives of geologists!

LSU

Upon completion of my 4 year duty in Jakarta, my wife and I decided to return to the US and complete my Ph.D. at LSU. I enrolled at LSU in 1989 and finished my degree in 1994. I had to have a Ph.D. if I ever planned to teach some day in an American university.

I made enough money in Indonesia to buy our house, cars, and pay for tuition and books upon my return to college. That gave us a big head start. But our everyday monies had to come largely from Jean's puppet shows and my monies earned teaching labs and lecture at LSU. It would be tight over the next 4-5 years. I was returning to school with four kids. Many considered that alone incredulous.

Interestingly enough, the home we purchased was directly across the street from the home of Dr. Arnold, a teacher who would become an advisor on my dissertation committee, so we thought this was meant to be. Arnold was a crass old man but possessed a heart of gold. I ran into Dr. Arnold many years later on a large tour bus as he, his wife, and I were slated to visit the Bingham Copper Mine as part of the AAPG (American Association of Petroleum Geologists) meeting in Salt Lake City. I asked how things were going and his wife began telling me his trials after Arnold was bitten by a brown recluse spider back home in Louisiana. He got very sick and was told by doctors that should he get bitten or stung by other creatures like shellfish or bee sting it could kill him. He was highly allergic. Hearing this and knowing I had seen bees swarming around a trash can earlier in the day I asked if Arnold was required to carry an epi-pen and his wife said "Oh yeah!" So I asked Arnold if he had his epi-pen with him and he said "No I left it in the goddam room." And he followed that with his classic response from a hard headed Dutchman, "I don't give a damn shit!"

Drs. Harry and Jim were great in the field also and I loved accompanying them on client-paid field trips to the lower delta. In fact one of Dr. Jim's favorite memories was the time I had to assist an over-weight woman geologist from the distributary mouth bar onto a rickety wooden pier by pushing her bottom up a few feet. That earned me the funniest memory, hands down, in Jim's book. Another interesting character was the Jewish isotope geologist, Dr. A., who came into class one day like he saw a ghost. Dr. A had entered an elevator and came face to face with three Arabs in full dress. He thought his life was over. Dr. A used to say "you people think that delta (as in delta-18 oxygen isotope) is an airline." We also reminisced about the time Dr. A. asked my friend Dave and me what was the point of our assigned reading assignment and after a few moments of groping with the task, Dr. A said "David…..and Cherry (for Gerry). Don't a bulla shitta me."

We loved our Baton Rouge neighborhood and we loved our home. And the Cajun people were just a hoot. Typifying the experience was a judge in town who regularly stopped by my house and asked my wife if I could come out to play. Another time, while walking down the road with this same judge we noticed a bag of garbage left by the side of the road. We sorted through the rubbish and found the address of the violator affixed to various envelopes. The judge then called this man and asked if it was his bag of garbage that was left on the roadside. Once he admitted it was his mess, the judge said he had 5 minutes to dispose of the trash or he would send him to jail. I like the idea of carrying a big stick. We also grew quite fond of two beignet (pr. *bane yay*) loving Cajuns and their two boys.

My first semester in grad school was brutal on all of us. You may remember we had six servants in Indonesia. And how many did we have in Baton Rouge? If you said zero you were correct. That meant Jean and I had to carry the whole load.

Jean made close friends with the women in the neighborhood and learned how to make red beans and rice, gumbo, fried catfish and King Cakes. And I noticed over our 4-year tenure, my blood cholesterol rising from 175 to over 220. In the two years following my departure from LSU, however, with a more normal diet, my cholesterol fell back to the 165-175 range. You are what you eat. And if you eat cracklings and boudin (pr. *boo-dan*) regularly your heart may just freeze up as well. In a related story, an excessively obese football coach reportedly stopped by a convenience store and asked how he could he get back to 335 (Route 335) and the female attendant reportedly said "Try eating salads."

Football at LSU was an amazing experience. Our home was located approximately a mile away from the stadium as the crow flies and I remember the roar and the ground roll that marked each LSU touchdown. In fact, a seismograph record hung in the geological sciences hall, only 200 or so meters from the stadium. It recorded a seismic event of considerable intensity and duration. Beneath the record is the title "Winning Touchdown vs. Auburn, 1988", a predictable earth response to 90,000 stomping feet. "Hold that Tiger" is the LSU fight song.

A very sad event marked our stay in Baton Rouge. It was Halloween and many people were dressed in costumes and on their ways to parties. With this background, a Japanese student dressed crazily arrived at the wrong house for a Halloween party. He rang the doorbell and the wife came to the door, peeped out and saw a dark skinned man in a scary outfit. She told her husband to get his gun. The Halloween visitor at this point turned to walk away when the husband opened the door and confronted him. He said "freeze" but this Japanese student had no idea what was the meaning of freeze, thinking he was referring to the temperature at which water turns solid. Tragically the man fired and killed this young man, setting up an international crisis between the US and Japan.

I spent 11 years full time in college (B.S., M.S., and Ph.D.) And the big question everyone asks is whether the tassel was worth the hassle? Economically, I imagine the answer is no. Folks with Masters Degrees are paid essentially the same as their doctoral colleagues. But the doctoral degree opens doors to high level jobs and opportunities for research and publication. Once inside, you must fight to stay, like everyone else. But you never have to apologize for not having the credentials if you go all the way. I'm glad I did it.

Argonne National Laboratory

Our family moved to Naperville, Illinois in 1994 where I spent 1.5 years at Argonne National Laboratory (ANL) in a post-doctoral fellowship. This was done to improve my resume and earn a teaching position at a recognized graduate institution. But my quest was never fulfilled, largely due to affirmative action mandates. At that time, universities had to make up for years of having no women in the departments by hiring only women. The closest I came to earning a tenure track position was my Candidacy with WMU. In it, I received notice I was in the final three in their selection process. And about two days later I received another memo stating they were sorry to inform me the position went to Kathleen D, ABD. Yes, the winning candidate was not yet degreed (ABD stands for *All But Dissertation*). This was maddening because all runners-up were told they failed because they lacked experience. The critical issue, however, was the sex of the applicant. My old teacher Dr. Will once confided that the search committee at WIU did not select me as a professor because they were told the winner must be a woman. I guess I was born at the wrong time.

One story of the kids, at this time, deserves mention. During the time I was working at Argonne, my wife and I drove the kids in our full-sized van to Yosemite Park, California. It was hot, as I remember, and cars passing us in the park were honking their horns. I did not know why I was being honked at until I stopped the car and went to the rear door and read this note written on the dusty window: *Honk at us. We are vacationing idiots!!* The note was planted by Dance and Sanny. We had a good laugh.

Visiting Lectureships

At this point and for the next 4 years, I began teaching as a Visiting Lecturer at a number of Chicago-based universities and loved it. I taught at Northeastern Illinois, Loyola, University of Illinois, Moraine Valley, and was awarded the Best Science

Instructor at one of the best universities in the country, Northwestern University (Evanston). But again I faced the reality of keeping the family of six solvent with an educator's income. So again I returned to Houston's oil business in pursuit of petro bucks.

BX

Disillusioned with not landing a tenure track position, I accepted a position with an oil service company by the name of BX and moved back to Houston. This, of course, meant I was temporarily giving up my dream of teaching at a research institution. My answer to all who asked was "the kids need shoes!"

BX hired me to be their Sedimentologist and one of perhaps six or seven company image log analysts, worldwide. Interpretation is never without controversy, however, and in one particular review, an email was sent that all peers come to a meeting to discuss my interpretation. A salesman remarked "But Dr. Kuecher has no peers." That felt good. I published several articles with BX including such topics as thin beds, turbidites, and wire line pressure correlation.

BX, at this time, asked me to accept a week's teaching assignment with CX in Angola, West Africa. Those who arranged the trip told me I should look for a man holding a CX sign on the airport tarmac. Furthermore I was warned against accepting a ride with anyone else but the man holding the CX sign. So the plane landed and I looked for a man holding the CX sign. But there was no man holding a CX sign. And as the crowd slowly dissipated, I realized I was the only white person not picked up at this airport, an airport known for ransom and kidnappings in the 1980's and 1990's. So I waited and waited. Nothing! I did not have the office telephone number nor did the phones work at the airport. And I came to the conclusion I must take a taxi or risk an overnight in Luanda International with hundreds of needy people hassling me to buy something from them, or worse. Anyway, after about an hour wait, I elected to take a taxi. The taxi driver I chose at least knew where the

CX office was located. And when I arrived safely at the office and met the host, I advised him no one picked me up and in a sort of epiphany this fellow said, "Oh, our sign needs to be updated. The fellow at the airport carried a sign that said Cabinda Gulf. But you of course were looking for sign that said CX. We were purchased by CX but nobody fixed the sign." That mistake could have cost me my life.

In another trip, I was sent to Uganda in East Africa to attend a field trip to Lake Albert Rift and teach locals from the area how to explore in it. Now just a few years previous, this country suffered a most tragic genocide and one could tell their souls were running on empty. There was little joy. The movie Hotel Rwanda revealed the power of hate radio to mobilize people to a mob mentality. As part of this trip we took a small safari trek to Murchison Falls to see crocodiles, hippos, and water buffalos jockeying for room on the shoreline. This was not a comforting development since we were placed in a dugout canoe with only 3 inches freeboard and paddled within 30 meters of this dangerous shoreline. These animals could have easily tipped our boat and pounded us into the mud.

We Lose John to Cancer

My brother John began suffering abdominal pain in April, 2002. He made visits to the doctor but the source of pain proved non-specific and difficult to identify. Doctors advised him he would have to deal with this pain, as if it were a stomach ache. But as time went on he felt worse and worse and he took more and more pain medicine to make it through the day. And in September, 2002, he was finally diagnosed with pancreatic cancer. Surgeons advised him that his cancer was inoperable so he began searching for alternatives. At this time he and his wife found a clinic in Houston that provided some hope and they made plans to get treatment. They stayed at our home in Houston the last week of December, 2002 and the first week of January, 2003 but the treatment didn't work because the medicine proved impossible to swallow and ineffective when he did swallow.

So he and his wife Sharon returned to their Frankfort, Illinois home and commenced a chemotherapy regimen.

By late September, 2003, he was around 100 pounds (down from his healthy weight of 250) and losing strength fast. And on the morning of October 24, 2003 he passed from this world at the age of 56. He was the first of Bob and Babe's children to pass away.

Our family was struck by the apparent randomness of the disease. Why John? Why now? We were all at a loss. His wife believed PCBs showered him and his boss in a work-related accident years before. Others suggested John may have ingested high doses of PCBs by eating contaminated coho salmon from the polluted Waukegan Harbor on the western shore of Lake Michigan.

KX in Kazakhstan

KX, a consulting company in Houston sent me on a 9 month assignment to Almaty, Kazakhstan in 2006. The first question you may ask might be "Just where in the hell is Kazakhstan?" Well Kazakhstan is in the western part of the old Soviet Union. Surprisingly, many of the native people, when asked, may tell you things were better under the old Soviet rule. This certainly was the case if the people questioned once had cushy political jobs. These newly freed countries were not regulated, and governed under an oligarchic form of capitalism where just a few people made all the money.

The operator for whom we consulted (XM) suffered a blowout drilling their first well and oil and deadly hydrogen sulfide gas was spilled. That alone was sufficient reason to expel XM from the country, but they survived and were drilling a second well when I exited the country.

I recently phoned a geologist who was with me in Kazakhstan and he advised me a scandal racked the operator company in

Kazakhstan soon after I left. Apparently, several top people violated the stock options part of their contracts. The operating company's CFO then departed hastily and the stock price fell. I remember most clearly the talk of lavish wine tasting parties held for the oil ministers. A few bottles reportedly cost 6,000 dollars each. Another area of concern was the cost of renting homes from a few select landlords in Almaty. This was the oil biz like it was years ago.

The most remarkable sights in this area of Kazakhstan are the ornate Russian Orthodox churches. When the Red Army over-ran the area in 1919, locals ran to the churches and carried away hundreds of priceless icons. The Red Army found only empty buildings and used the churches to house horses. But when the Soviets Union collapsed, the buildings and domes were restored and the relics returned. These icons survived over 90 years in the basements of believers. What a testimony!

Saudi Arabia, Aramco

I decided, in 2007, it would be wonderful to have worked the greatest oil fields in the world and shortly thereafter accepted an overseas position with Saudi Aramco in Saudi Arabia. Getting the offer was the easy part. That only took a few days. The most time consuming thing to do was complete the paperwork, the physical, and the clearances. It took 120 days from the day I accepted the offer to boots on the ground in Saudi. Heck I could have found 10 jobs in that time! And I kissed my wife and exchanged my home in verdant Houston for a place so foreign, hot, and dusty that it could have been the moon. Saudi Aramco recruiters continue to be baffled, for example, how they can make 15 competitive offers and get just two people to sign on. Quality of life to Aramco is money and that is not the only consideration for young geoscientists today.

I remember meeting a representative of the company at the airport and, upon leaving the doors of the terminal, stating "It doesn't seem so hot here" and he replied "That's why they bring you in at

2:00 AM!!" And I knew I was in for a big surprise! I would have to grow tolerant of summer temperatures in the range 40-55 Centigrade (104-131 Fahrenheit), high winds and occasional shamal dust storms. In fairness, however, there are few places nicer than Saudi in the winter months.

Arabia is a hyper-arid desert that receives between 0.5 and 1.5 cm rain per year. The extreme low humidity translates to blue, cloudless, but occasionally dusty skies, generally from May to October. I was so used to blue skies, in fact, that when I saw my first cloud in mid-October I shrieked, "Look! Clouds!" Now that is no big deal in Texas.

Saudi Arabia is 8 time zones ahead of Houston. I dealt with that difference by simple arithmetic. Jean, however, had a more graphic solution. She was aware our family members lived in a number of different time zones (Dance lived in New York, Sanny lived in Chicago, Larue and Jean lived in Houston, Dane lived in Austin, and I lived in Saudi Arabia). So she purchased 6 large clocks and posted the local time at each person's locality. This display could be seen as one climbed the flight of stairs near the home entrance. This was a clever idea and made a statement of who we are as a family. And it explained why it was so weird getting up at 4:00 AM to watch a ballgame being played at 8:00 PM back in the States.

Saudi Arabia has a rich history. Petroglyphs (writings on stained rock outcrops) dated at some 4000 years old tell a story of a savannah rich with grasslands, game, and predators. These petroglyphs document a story of rapid climatic change. Saudi Arabia was not the desert we see today 4000 years ago. It was more like the grasslands of Tanzania. The Nabataeans occupied this land and built ornate stone burial grounds at Mada'in Saleh. Similar structures were built by the ancestors of these people at Petra in Jordan.

Saudi religious scholars for years had taken the position there was no history before Mohammad. This position encouraged the removal of Neolithic pre-Mohammad antiquities. Aramcons and others were of two positions on this issue, namely: 1) Take what you want because someone else will if you pass this opportunity, or 2) It is better you remove the Saudi artifact than someone who does not appreciate it. The outcome of this confusion resulted in Aramcons removing tons of pottery, shards, petrified wood, arrow heads, spear points and fossils. In fact, visiting an Aramcon home typically involves a tour of that family's bounty. There is a new movement afoot, however, to ask Aramcons and others possessing antiquities to return them. I think this is great, although as they say in America "the horse had already left the barn."

Saudi Arabia's king has the responsibility to protect the Holy cities of Mecca and Medina, the spiritual home to one and a half billion Muslims. And although King Abdullah has been a hard working element for change, the western world fears a fundamentalist could seize power in Abdullah's absence, and set the country back 50 years or more.

The King of Saudi Arabia draws a lot of fire from fundamentalists, especially when the king asks to reconsider fatwas like allowing women to drive or permitting movies to be shown in areas outside the Aramco community. The King also provides meeting places for Christian ex-pats on the Aramco campus. It is technically illegal to build a church on camp or anywhere in Saudi Arabia but it is not illegal for an expat to attend a service in an on-campus location.

The greatest future drain on the economy will likely be princely entitlements. Every male born of the king's lineage has an incredible salary promised to him for life. That, of course, encourages princes to raise very large families. In fact, if the number of princes continues to grow at the current rate there may be a tipping point in the near future. It's like the math question that follows:

Would you rather have a penny doubled every day over a 30 day month or a million dollars cash up front? On the surface it looks like the million dollars up front is the better deal but if you work out the math, the doubling choice is best. This is a very similar problem as the growing number of princes and their entitlements. There are thousands of princes (current estimate around 5500) and the number doubles at an alarming rate. At some point there will be a crisis where more entitlement money is going out to the princes than money coming in to the country. What may save the Kingdom, however, may be the discovery of new oil and gas reservoirs in the Red Sea, shale gas deep in the Arabian Basin, and other exploration strategies yet to be proven.

The religious police, called the mutawas, sport long gray beards and wear thobes (male Arabic dress) with unusual "high water" pant legs. They may also carry a thin stick for smacking at the violator's ankles, specifically the level of young ladies skirts. We often called mutawas "the fashion police." You can usually find mutawas at the malls where they zero in on young people. Expats keep some distance from mutawas, who may or may not be accompanied by local police. It is important to note that many of the infractions are violations against tribal rules and not violations against the Quran.

Needless to say, most women have it rough in the Kingdom. Aramcons call Saudi Arabia *The Magic Kingdom.* There is no telling how many bizarre fatwas (opinions to questions in life by mufti holy men) there are on the local law books. Fatwas are largely designed for the benefit of men. We are all likely aware of the fatwas against women drivers and the fatwas requiring women wear full-length abayas. Additionally, women cannot exit the country without the written permission of the husband. There are many more bizarre and troubling fatwas on the books in Saudi.

Many western folk have difficulties accepting the Saudi arranged marriage. We find it unusual that Saudi parents still, in 2013, find the spouse for their children. The children accept the selection of their mate out of respect for the parents. But unlike western marriages where young people marry each other, Saudi couples marry the families. In this way the Saudi marriage is different from what we know and may have value to modern marriages. The best marriages occur when both sides like each other.

Saudi Arabia has a different work schedule than the rest of the world. Saudi weekend days are Thursdays and Fridays and of course the US works Mondays through Fridays. So when Saudi opens for business on Saturdays, business transactions are typically delayed until the rest of the world returns to work on Mondays. Most recently (June 24, 2013), however, the Saudi King followed Bahrain's lead and announced Saudi weekends would commence on Fridays. Now Saudi is only one day out of phase with the rest of the world.

Businesses, additionally, must close four times per day for prayer and the timings of these closures are posted daily or reviewed on mobile phone apps. The important concept here is one cannot just go shopping expecting a business to be open, like in the west. The key is to arrive before the doors close. Workers will commonly allow you to finish your shopping if you are inside the store at the time of closure.

Saudis often get paid a salary to go to an overseas university, especially if they are supported by Saudi Aramco or some other business in Saudi. Having a generous income in graduate school, however, generates some bizarre situations, like what happened at Colorado School of Mines geology field camp a few years ago. Field camp traditionally means roughing it in tents. But the Saudi students purchased a motorized camping trailer for seven students that included a cable TV, a refrigerator, air conditioning and a prayer room. This did not go well with CSM instructors and fellow students, however, and a policy change was enacted. Field camp is an experience not just an inconvenience. Every geologist should experience the camp part of field camp.

Productivity in Saudi Arabia falls off dramatically during the holy month of Ramadan. But if Saudi Arabia can produce a barrel of oil for $4.50 per barrel (the lifting cost) then the western concept of productivity is thrown a curve ball. This place is swimming in oil. In fact, the oil productive area in the Eastern Province alone is larger than the Netherlands and expanding every year with new discoveries. In fact, a gallon of gasoline was only about 27 cents per gallon in the years I lived in Saudi Arabia. Saudis, in response, generally purchased large vehicles and drove with reckless impunity.

Expat housing is provided by Aramco in various work camps. These homes are simple in design but comfortable. There are four main communities or camps. Dhahran is the largest camp with an expat population of about 11,300, Ras Tanura is next largest with about 3,200, followed by Abqaiq with 1,950 and Udhailiyah with 1,350 (Aramco magazine: The Power to Provide). The total non-native work force, however, is much larger, perhaps 6 million.

Expats love to travel to nearby countries because they may never again be so close. Yet there is also a weariness that develops when ex-pats both live and vacation in the Muslim world. I would argue it's good to take the opportunity to return home once or twice a year to refresh oneself in western culture. Aramco is familiar with this weariness and encourages expatriates to take at least one lengthy vacation (re-pat) home.

Saudi Arabia is not open to tourism. Only your immediate family can visit you in Saudi Arabia. Guests must have valid visas and be sponsored by an Aramcon.

Expatriate living develops strong bonds. The mantra "We are all in the same boat now" prevails and ex-pats cling to friends like they were family. In fact, these folks are often the only ones you can relate to over the years because your experience is just too foreign and strange to share with friends back home.

Many in the ex-pat community refer to Saudi as "the logic-free zone." In support of that, I was once in an accident in which a local backed in to my driver's side door, and after the dust had settled at a local police station, the police issued me the ticket stating there would not have been an accident had I not been in the country.

CX's imprint is all over Aramco. Their most important contributions included the taking of conventional core for every vertical (pilot) hole and water pressure maintenance from the very beginning. It is also a CX legacy to set up camps for workers, and to have a different set of rules within and outside the walls of these camps. The place is just like Mayberry. Heck I found myself whistling that tune on several occasions.

My friends, Mk and Bl, once asked me why I came to Saudi and I responded "To make a difference" and they shrieked "That was your first mistake!" "This place", they explained, "has a plantation mentality" and the two of them went directly into a stand-up comedy routine "Yessa boss, I'll pick you cotton." We are, essentially, the hired help. And if there is any doubt what country watches over Saudi Arabia, all one has to do is walk outside in the late afternoon and experience the ear splitting roar of four or five American-made F-15's flying low on training missions in a show of force and camaraderie. Strange bedfellows the Saudis and the Americans!

Saudis want to work in air-conditioned office environments. And in a few generations, the Saudis appear to have lost their abilities to live in the desert like their Bedouin ancestors. Saudis prefer indoor labor and outsource outdoor labor. Furthermore, Saudis want to manage projects regardless if they have the technical experience to handle such a job. Instead, they surround themselves with expatriate competence in hope they can cruise to managerial knowledge without first gaining the required on the job training.

Recently, Aramco offered buyouts to retire up to 5000 Saudi middle managers in an effort to open the promotional bottleneck.

These middle managers reached a stall point in their professional growth. Many Aramco senior managers believe the current generation of professional Saudis do not share the hard work ethic of generations past. But these young people like Aramco because of the great benefits in working there and their abilities to walk from their jobs at 4:00 PM. This is not how business in the US operates.

Apparently, the worst thing you can do in Saudi Arabia is to embarrass your boss or fellow worker in front of others. Locally, they call this *losing face*. Many an employee has been dismissed for embarrassing their bosses in public. You learn to always provide wiggle room for the person in question to escape, even if that person is non-productive. But the worst thing about this system is you never know who will be offended. You may have said something years ago and when it came to your evaluation you were clueless why your employee reviews did not make sense with all you were told by your immediate supervisors. Maybe a senior VP overheard an incriminating conversation while walking the halls. You may never know the real reason. Aramco supervisors rarely confront. It's all rather covert.

Unemployment is high in Saudi Arabia and that is a concern for stability in the country. In an effort to reduce unemployment, Saudi Aramco decided to build the world's largest petrochemical refinery. The new facility promised to provide thousands of jobs for construction workers. But construction jobs are relegated largely to Pakistani, Filipino, and Bangladeshi foreign laborers because Saudis don't typically work outside, let alone, work construction jobs. It's too hot. And with these plants becoming more and more computer-automated with time, there are very few jobs created considering the massive capital investment. These projects do not significantly reduce Saudi unemployment. Most affected by this disillusionment are the Shia because the Sunnis reserve office jobs for themselves. A sense of distrust between the Shia and the Sunni is palpable.

Security is a constant threat outside the Aramco camp and the US

Consulate issues warnings regularly. Fifteen of the nineteen 9-11 terrorists originated from Arabia, and Saudi police, thereafter, rounded up the usual suspects, some say over a thousand. Most of those were not heard from again.

It is not a good idea to travel to distant parts of the peninsula un-escorted by Security Police. You must have identity papers for every trip. These precautions are in place because Islamic terrorists killed 22 expats in 2004 at the Oasis compound, just outside the Aramco gates. Locals call this event the Al-Khobar massacre and it goes to show you can never completely let your guard down, no matter how safe it appears.

My friends Mk and Bl openly discussed evacuation plans with me should things go badly in Saudi. We decided to pack canned goods, bottled water, our passports and lots of money in backpacks so we could grab them and leave in a hurry, preferably to Bahrain. The plan was to eat all the goods in the backpack and re-pack with fresh food every six months. Bl had a 4-WD all-terrain vehicle but the exact evacuation route was dependent on the news of the day.

The key thing to remember, when Americans and Saudis measure each other is that we are ambassadors of our own cultures. We each have a duty to act responsibly and respectfully. One thing the average Saudi could clean up is the practice of clearing their noses in public washroom sinks. Likewise, ex-pats should refrain from pointing with their fingers, showing the bottoms of their shoes, raising their voice in conversations, shaking the hands of women, and failing to wash your hands after using the washroom, among other things. Your Arab brothers and sisters are watching you. Be good ambassadors.

Shopping is an obsession in Saudi and on holidays, everybody makes their way to the suq markets. My favorite shop in the old Al Khobar suq was Roma Tailor. The fellows in that shop knew me by name. I had all my shirts tailored at Roma's, new shirts made, and more importantly had a beautiful tuxedo custom made just for me. In fact, Roma Tailor had a fashion catalog and I was a featured model. Check it out!

I was on Bachelor Status in my time at Aramco but I was not really a bachelor. No, I was happily married to my wife back home in the US. We did have one advantage over those on Married Status, however. We earned 50 days holiday vacation while my married colleagues only got 38, the difference being 1 vacation day per month. The married folk got even, however, with full fare plane rides for the whole family. Several of my friends with substantial families hauled in paydays well in excess of 30,000 dollars for annual re-pat benefits. Young families, in this sense, are Aramco's happiest transplants.

I missed home and ended up hanging at the home of a nice geophysicist named Pt and his wife. I loved playing with their kids. Their youngest daughter was a real character. I remember telling her mother how quiet she was when I first met her and amazed at how outgoing she had become. And this young child, standing next to her mom exclaimed "That's when I was sick." I have not forgotten those kids and occasionally send them PEZ candies from the US.

Pork is illegal in the Kingdom so you must be very careful if you transport it from Bahrain to Saudi Arabia. There is a major custom stop on the Causeway and custom officials often ask if you are carrying pork. On one visit, my friend Mk told an official he was not carrying pork. He was, instead carrying bacon. Are you sure? Oh yes, it is not pork, it is bacon. Surely some of the guards knew that pork was bacon but laughed because Mk was so clever and let him pass. That's how it is there.

The same rules hold true for alcohol. Alcohol is technically illegal to own or drink but everybody knows when Commissary shoppers buy sugar and grape juice in bulk they are intending to brew some hooch. In fact, Aramcons openly engage in wine, beer, and whisky tasting competitions. The local wines were not bad. The brown whiskey, locally known as *sidiki*, however, tasted like poison. The American Consulate, considered US territory, served hard liquor at their parties.

It is likewise technically unlawful to create or attend puppet shows in Saudi Arabia. So when my wife Jean (a puppeteer) visited me in Saudi, a clandestine operation was set in place at the British, American, and the Arab schools to permit these children to see western theatre. I was tickled when Jean re-named the children's story "The three pigs" to "The three monkeys" specifically for Saudi audiences. Censorship extends to the written word and imagery, and is both maddening and amusing to see large black-out portions of newspapers and magazines.

Saudis have developed an obsession for sweets and fried foods and line up at our western franchises like Cinnabon, Dunkin Donuts, and Kentucky Fried Chicken. Needless to say, obesity, heart disease, and diabetes are on the rise in the Magic Kingdom.

Jean's Memorable Trip to Saudi

One evening Jean took a few minutes to send an email to the kids discussing her trip to Saudi. It was an exhaustive journey and well worth the read:

I finally arrived in Saudi.
It was a difficult journey:
Houston to Dallas was OK.
Dallas to NYC: Flights canceled due to snowstorm. Arrived 4 hours late but in time to go to a play.
NYC to London: Plane replaced due to mechanical problems, arrived 3 hours late, missed nonstop to Bahrain. Would have arrived in Bahrain Monday night but 5 hour layover in London and was re-routed.
Flew London to Abu Dhabi: Terrible turbulence. Arrived at

12:30 AM
  Flew Abu Dhabi to Bahrain, Got on plane at 2:30 AM in Abu Dhabi, Giant thunderstorm came while we are on runway. Worst rain storm in 20 years. Plane sits on ground but moves in the wind. Will not fly.
  Unload from plane on outdoor stairs into buses.
  First floor of airport is flooded. Wait in lines for new boarding passes.
  Got a meal voucher, no place to sleep.
  Borrowed a cell phone of South African woman to call Gerry.
  Finally depart on another plane at 8:45 AM.
  Arrive in Bahrain Tuesday morning at 9:30 AM.
  Gerry picks me up. Stop at Aramco personnel office to apply for visitor ID card.
  Home, ate half of fruit cake for lunch, Gerry goes back to work. Time for a nap!
  All that rain in Abu Dhabi and nothing here. The signs and plants are dusty.
  Dad is fine,
  All is well here. Hope you are all well. Love, Mom

Jean's Friend Responds to Her Note Above

Girl,
  You are one of the most phenomenal women in my life. What struggles you graciously overcame to get across the world! You are a biblical wife. Your support for your husband and family always inspires. Your kind heart towards others, your modeling of patience and your intelligent fortitude in times of trouble are setting examples for others here and across the world. I am blessed to call you friend. Please keep me in your family loop with e-mails. I feel like I am there with you seeing weather, talking w/ vendors and walking streets while reading your words. I am so glad that you are safely on the ground.... no matter how foreign the customs I trust you will keep all the strange beliefs in perspective. You are where God wants you to be. Love you, my friend.

Exiting the Kingdom

Pt sent me an email when he learned I would be leaving Aramco that went like this: *"We have been so blessed by your time with us here. You've been 'Uncle Gerry' to our kids and a great friend to me. I'm really going to miss you. But I'm sure we will keep in touch. And maybe it is time to go home. You've had an interesting career. You have blessed many just by being yourself. You are a wonderful combination of smart and easy-going friendly that is really rare."*

I found it more difficult to tell my friend Mk, who counted on me for support. My email to him went as so: *"I will be leaving KSA this August 10 on a one way flight to Houston. And although I have been around the world in my career, I won't get around much anymore." Heck, I may as well purchase a square toe kicking shoe and try out for a college football team!*

Gerry's Retirement Speech, July 6, 2011

This is the speech I gave at my Retirement Breakfast:

I look around and I think there must be some mistake. Are you sure we aren't celebrating someone else's retirement? I'm too young for this. They're "putting me out to pasture", as they say on the farm. At least that's better than the Old West solution "take him out and shoot him".

I want to thank everyone who came today and advise all of you that the list for this event was carefully drawn so as to not include persons to whom I may owe money.

I spent my whole life preparing to work and now I'm preparing to not work. That is weird.

My life as a professional geologist has been a long and winding road and I have seen a lot. But I must share with you the first words I learned in Arabic I learned in Egypt. It didn't take long either. In fact I learned these words within 15 minutes of my arrival at the Cairo airport. Those words were bakhsheesh (meaning tip or bribe) and bukrah (meaning later or tomorrow).

I want to digress to the day I left home for this assignment, June 14, 2007. I remember security in Houston found a small pocket knife in my bag and took it away. Now, some four years and three months later I'm leaving Arabia and I may be carrying a sword, if that's what I think is in that box.

I made a number of excursions into the desert at Shaybah, Qaisum, Wijh, Jiddah, Hail, Mada'in Saleh, Dubai, and into Oman and each time was amused by the stewardess of the flight blindly remarking "In case of a water landing" What water?

So I have been here 4 years and 3 months and I am retiring. I have been gone from my family a long time and there is much to catch up on. Over the years my opening line to my wife as I came in the door has changed from "Honey I'm home" to "Honey I'm home again" to "Honey I'm home forever."

I'm sure there will be adjustments when I return home. In a previous assignment (Indonesia), we had 6 servants. Currently we have none so the solution involves me.

I was retired from Saudi Aramco when I reached the age of 60, a mandatory company policy. Leaving Saudi involved selling all the things that made my little Dhahran apartment a home. One of the things I sold on leaving was a queen size bed. I had plenty of room rolling about on the queen size bed. When I sold it I was forced to use a small single bed from Aramco Furniture, just to get through the last month. Well it only took me two days to roll off that small bed and bust my eye on the night stand. My wife correctly assessed the situation and said I should move the night stand a few feet further from my bed. My mother, age 87, innocently offered her own solution. She said why don't you hire someone to sleep with you?

An Arab friend, Muf, once shared a dream he had of me. In the dream he asked a friend if he had seen me and the friend said he heard I boarded a kayak and was paddling to North America. "Let me out now" comes through pretty clearly, eh!

My time in KSA was rewarding. My first year was spent in Geosteering and the last three years in training. I taught three classes in my tenure, namely Geology for Non-Geologists, Petroleum Geology, and Borehole Imaging. Overall, some 350 Saudi students had a week's training with me. In addition I mentored students at KFUPM (with the help of a KFUPM professor) to compete in the 2009 American Association of Petroleum Geologists International Barrel Award. These young people won the Mideast Regional First Prize and placed Third globally at the AAPG in New Orleans. Additionally I conducted, with the help of Dr. GG, a seminar at KFUPM entitled "How to get and keep a geoscience job" which focused of effective resume writing and interview skills. In summary, I think I made an impact on the incoming Saudi young people, the next generation of Aramco employees.

My family has chased me around for years at a number of overseas and domestic assignments and it's time to repay them for their loyalty. I need to be home although the concept of just sitting around sounds like a recipe for a heart attack. So I will continue to consult and teach. It would be a shame not to continue, after all the years of preparation to work (11 years of college). Please note my website geraldkuecher.com and keep in touch. I'll always remember a most simple workplace joy. I hung a prism in my office window that cast a lovely spectrum in the early morning. Folks coming for coffee loved it and so did I. They claimed the beam gave them a charge for the day.

Retirees often say every day is Saturday or in this culture, every day is Friday. And maybe I will feel that way as well. But I have always been a hard worker and cannot imagine being idle. I should land on my feet

But my most important message concerns how I have been blessed to know the people in this room. I really felt I made inroads with my Arab brothers and colleagues and was invited on many occasions to share meals, dig for sand rose crystals, go on field trips, purchase shrimp at local fish markets, visit muscle car conventions, visit Bahrain, and even over-night at Saudi homes. This was my personal legacy: to bridge the cultural gap. The world changes one heart at a time.

Likewise I found great comfort in my Aramcon friends who have helped me at every step and have valued my friendship more than any other colleagues I have ever known. And for this I extend a great big thank you. And I thank my wife who made my life as an itinerant geologist possible. It has been a great ride.

I opened my remarks with the first two words I learned in Arabic. Now I want to come full circle on that opening thought with two words I think appropriately summarize my feelings at this time and these words are shoukhran and maasalama meaning thank you and good bye, respectively.

The Rush to Leave the Kingdom

About two weeks after my Retirement Breakfast I lost my security ID. This is a major no-no at Aramco because they stop you so regularly at check points. I tried twice to re-trace my steps, but to no avail. I double-checked home, my office, my pockets, my car and I had no luck. So I called my remarkable mother, Babe. She listened to my situation and said she would pray to St. Anthony, the patron saint of lost items. Don't ask how St. Anthony ever earned such a reputation but this stuff really works! Within two minutes of Babe praying, I had a knock on the front door. It was my ex-house boy. Incredibly, he said he had my ID. "Where did you find it?" I asked! He said he found it a atop a garbage bag in the trash bin. He surmised I threw the ID away as I pitched my own garbage in the bin. Or perhaps he was going through the garbage. I don't know. But it was a clear answer to prayer.

This was a miracle. Babe's hot line to St. Anthony saved me a fee of between 500 and 1000 USD. I called Babe back on two occasions that same night thanking her and advised her that this miracle may just end up making its mark around here. My friend Mk said he plans to call Babe the next time he loses something. This thing may spread.

St. Joseph is another mystery. Catholics believe if you bury a statue of St. Joseph upside down and facing your home, then the house will sell. If however, the statue points in another direction then perhaps you will sell the neighbor's house, intended or not.

In my final 5 weeks in the Magic Kingdom I taught three week-long classes and was under considerable pressure to wrap up all loose ends. So one weekend I decided I would cross the Causeway Bridge to Bahrain with some friends. And in one of the clubs, I encountered a sight I will not quickly forget. The bar was loaded with Saudis in thobes drinking like the most seasoned Irishmen!

I expected to see in Saudi Arabia, an occasional young man with just one hand (Sharia punishment for stealing). But I never saw that in all my time. I think this is a law on the books but seldom, if ever, used. There is usually someone in the family with sufficient wasta (influence) to escape such brandings. After all, most petty thievery is by young children. Beheadings for adults, however, still occur as nearby as the IKEA store parking lot in the Dammam Mall. And isn't it funny that hand choppings are seldom employed but head choppings are still practiced. Apparently there are less political fireworks if the person is eliminated.

Arab Spring recently swept the Mediterranean borderlands. At first the political cause was thought to be instigated by Ismamic extremists. But as the revolution spread from Libya, Tunisia, Bahrain, Saudi Arabia, Syria, Egypt, and elsewhere, the issue became clear it was not anti-American nor was it pro-extremist. Rather the issue was anti-corruption. Over the years, these leaders had stolen billions and the people grew sick and tired of corrupt, tyrannical rule. Westerners should take comfort, I believe, in the rebel smiles seen on international news, the troops nudging forward to get themselves photographed. Arab Spring, I believe, will likely issue in a new era of openness and pride for these countries and a demand for financial accountability. Power corrupts, however, and governance can easily return to its old rotten ways.

The most difficult thing to understand regarding the Saudi response to unrest in Bahrain was the arrest of doctors and nurses who patched up injured demonstrators. Hello! Attending to injuries is the Hippocratic duty of doctors and nurses. Doctors are obligated to respond, regardless of the political affiliation of the injured.

Considering the vast number of students sent overseas to study by a number of these Arab governments, it is quite likely that many of these students return to their countries with markedly different values from those values left behind. We cannot under-estimate the effect of cell phones and internet on opening the world. Seems like every Saudi has a cell phone and regularly uses it. This is having a profound effect on the westernization of Saudi Arabia.

Folks back home wonder what was my take on the Muslim faith, having lived some 8.5 years in Muslim Indonesia, Saudi Arabia, and Kazakhstan. And although I have no right to compare great religions, I would argue I grew in tolerance and came to appreciate the Muslim heritage as well as finding a renewed conviction for my own western traditions. *We just need to learn to live together* (famous quote of Rodney King whose televised beating started the Los Angeles riots). It is important to note that the Old Testament is embraced both by Christians and Muslims. Additionally both Christians and Muslims believe Christ (Jesus) will come again.

In summary, the Arab people are prayerful people and courteous co-workers. And most of them worked well with their American colleagues.

Author's goodbye "maasalama" embrace
with close Saudi friend

Bob's Last Years

Bob came face to face with the prospect of an early departure from Earth at age 56 when he suffered a major heart attack. Bob recovered, however, and continued to work in the shop. But by the late 1990s (as best as Babe remembers), he began to reduce his shop time from 10 to maybe 4 hours per day. Babe called Bob every hour or so to see if he was alright and that seemed to work. It was a bit awkward having him in the house for such lengths of time, however, especially since we had never seen him just hang around. He was a working man's man.

Bob experienced a stroke in his right eye at age 82. Bob said he was sitting in a chair and suddenly, as he remembered, "the lights went out." That eye would never again be useful.

Bob's overall health slowly worsened over time and he resigned his role as the capable one and replaced that with the role of friend and counsel. But the good news was that Bob became available to his family in a more effective way for the first time and began saying he loved us for the first time. Talk about awkward!

Sanny, our oldest, had a special relationship with her grandfather. He appreciated her sharp wit and her passion and respected her immensely, writing her a number of letters. She knew how to impress him as well, buying him an occasional bottle of Crown Royal.

A letter Bob wrote to Sanny reveals the love he had for life and the mischief in the mind of an old man whose time had passed. The note, translated from Bob's hieroglyphics follows:

April 29, 2008

Greetings from Palos Hills, better known as Heaven on Earth, and congratulations on becoming a lawyer. We'll celebrate my birthday May 9 when you arrive. We will celebrate with my treasure, Crown Royal. I suggest you don't dilly en-route as the bottle will be getting near the bottom. I hate to think about those things. It ruins my day.

Anna (my aunt and your great great aunt) was a saloon keeper who was hard of hearing. Anna believed anything a good spender told her and ran the Tampier Saloon on Kean Avenue across the street from my home as a boy. She was too proud to get a hearing aid and didn't hear half the conversation anyway. So can you imagine the stories she would conjure and what she thought she heard. Occasionally Anna would visit my mother. I would listen to some of her stories and I'd laugh so hard that mother would send me outside.

Those were the days of the Roaring 20's and gangsters and the stories would make your hair curl. Al Capone, the Genna Brothers, Vince Mackerlain and more. Anna tended the bar. It was Prohibition and the booze and money flowed freely. One summer day a caravan of trucks came by our house. One of these trucks had a problem with its radiator. The driver asked me to get some water and I did. After a few minutes he brought back the water can and he gave me a 5 dollar tip. My father would work a week to get 5 dollars! Their trucks were carrying barrels of beer. Yes Sanny, I am a fugitive of the Roaring 20's. I experienced both famine and plenty. I witnessed two uncles die from alcohol abuse. I also saw men go blind from drinking wood alcohol. Incidentally Aunt Anna's husband also died from booze.

Grandma sends her love. So do I. The Queen just called from her boudoir-her voice has that come hither sound. I think she wishes to roll in the hay. Duty calls.

See you later. Grandpa Bob.

Bob wanted us to know he had lived a full life and had seen many things, some good and some bad. His family, specifically his Aunt Anna, was involved in Tony's booze joint at "the end of the road" where Kean Avenue meets the canal. Goes to show people will do what they have to do in order to survive.

Thinking back on the bar the locals called Tony's, one cannot escape the white elephant the bar represented. During Prohibition, everybody knew what was going on but no one said anything. Men often headed off to Tony's in the evening, their wives reportedly showing their displeasure by ignoring them the rest of the day.

Another issue in my childhood concerned the Mexican hacienda the turf nursery built for migrant Mexicans. Everyone knew illegal aliens were coming each year and staying for months and even years to work, yet nobody said a word about it. There was a fierce, anti-government independence shared by the Palos locals that enabled folks to make a living regardless if laws were broken in doing so. Not much has changed since Prohibition. Winston Churchill once said "Americans will eventually do the right thing after they have exhausted all other possibilities."

Bob was both a heroic and a tragic figure. To us kids he was the tall, strong man of the West, a bit like Chuck Connors of the TV series *The Rifleman*. He was always capable and self-assured. But unlike the Rifleman, there was little private time available or offered for gentle fatherly counsel. The kids suffered from this. He became a better father at the end of his life, however, and for this we praise God.

Bob loved his life atop the hill and loved looking out the bay windows at breakfast time for deer that would come in to eat acorns. When one of us sighted a deer, the family would silently find a comfortable spot and watch in amazement as the deer fed. Bob loved and honored the deer. He did not have the same feelings for coyotes, however. He felt they could kill and run off with one of our dogs, so Bob would shoot at them from his bedroom window.

Bob asked me in 2008 if I would like to invest in the shop. Initially I did not know what to make of the request and said no, I was not interested. On the surface, my monies could be better invested elsewhere. But that is not what Bob meant. Bob was asking for my help because a building had burned down and needed to be rebuilt. This didn't translate from his initial inquiry. My mother advised me several months later just what he meant and Jean and I gladly loaned them the money. Bob couldn't admit he needed the help. This to me was tragic because I would have come to his assistance much sooner had I known the real story. Communication was not his greatest skill.

Bob grew more and more incapacitated with lung cancer as time went on and had to be shaved, bathed, and occasionally helped in the bathroom. He was a proud man, however, and I am sure this was not easy. My sister, Carolyn, was his live-in nurse, administering his medicines and care 24-7. She did the family a great service. And as Bob slipped towards the end he called each of his children to his bed and shared a few thoughts. Ed claimed he said "Look at me, Ed. I am the face of death. Enjoy life." I don't know what he said to others but he asked me, as executor of his will, to keep the family together, and I pledged to keep the family together with occasional picnics and get-togethers.

In his last few days, Bob was seen pointing to the ceiling, as if to say "Do you see what I see?" Babe and Carolyn interpreted the action as Bob seeing a vision of his sister Gladys calling him home. Bob even spoke to Babe and said "I'll be leaving you soon" to which Babe asked "Just where do you think you're going?" Of course we all knew. There was no hiding the obvious.

On his last morning, Bob was having some trouble breathing and Carolyn called for an ambulance. And as he was loaded in the vehicle he had a look of resignation that this may be his last field trip. But he bravely went off.

The attending ER doctor spoke frankly to Bob and those in attendance that Bob's condition was grave and asked the nurse to administer morphine to *make him more comfortable*. Soon thereafter a priest was called to administer the Last Rites. Bob had expressed earlier he did not wish to have Last Rites but did not reject them when the priest arrived. In fact he thanked him. Bob was sitting up when my sister Janet noticed a slump of his head and he was gone.

The end for Bob didn't come as a surprise. He was diagnosed months previously with lung and facial cancer. It was a matter of time and he knew it. He smoked for some 40 years and although he gave up on the habit years ago, the damage was done. Carolyn took over hospice duties in my father's last months and he was fortunate to have her as his full time nurse.

Bob passed away on April 17, 2009, three weeks short of his 90th birthday, following a 2-year bout with lung cancer. I was notified he passed away while I was on vacation in the Netherlands and immediately returned home.

Bob was waked in the traditional way. The family, by this time, was emotionally and physically exhausted and prepared to say good bye. My sister Carolyn made all the arrangements including a police-escorted final drive around the property. Even Elvis (a Las Vegas style entertainer who worked for the City of Palos Hills) showed up and sang his tribute "You were always on my mind." Bob would have loved it. In the end there was music.

When one visits the house today and enters the family room they are drawn to a green leather arm chair in the room's distant reach. This is where Bob spent his remaining days reading newspapers and magazines. Today the chair is strangely empty. And his lap dog, PG, doesn't seem to know just what to do. So she joins the other dogs, taking their positions on one of the two arms of the sofa and stare out the window, doing their best to protect and defend. This is all they know, these vigilant sentinels.

On Bob's left was a light that turned on and off with a switch at its base. Remarkably this light turns on and off today as if it had its own mind. Babe and Carolyn would say this is Bob re-visiting them. "He comes most often when the family has a big decision to make" Babe would say. Maybe it's an electrical short. We don't know. But who are we to say it is not Bob? Maybe, just maybe, Bob is letting us know he is still with us. But to say all decisions made in our home were borne of logic and methodical pragmatism would not be the truth. Instead there are elements of paranormal activity interspersed within. The family, it appears, uses these experiences of the dead visiting them to support their agendas. And my sister who lives on site may say the light turned on last night and that's Bob telling us this or that. Case in point, the light came on recently when a limo drove up to the house to take Babe, Carolyn, and a few other girls to a special birthday party for Carolyn. What was Bob trying to convey? I'm sure it had to do with the limo and Carolyn spending all that money on her birthday. The light is especially active when we sip Crown Royal whiskey in Bob's honor. Salut!!!

Bob was a powerful personality, to be sure, and it is not surprising he comes back in our memories. I recently heard a haunting tune while driving through the deserts of Saudi Arabia. I knew I had heard it before but could not remember the words. And as I rounded a turn, I saw an area of intense blackness and then saw Bob's softly lit hands on his saxophone delicately playing a tune. It was like a well-choreographed black and white movie. Later I recalled the name of the tune, *Stranger on the Shore.* Bob played this tune so well. It is clear that Bob and his music are still with us.

I dream of my Dad and I dream of my brother John. I smile and cast them an occasional wink. John often comes to me in visions where he dons a top hat, a stylish departure from his many years as a poorly dressed young man. Why a top hat? I don't know. And I think of my Dad sitting in his green leather arm chair surrounded by his dogs and a newspaper and I smack my leg in disgust that I did not come to his aid sooner when asked for a loan.

I do believe the greatest lesson I learned from my father, Bob, was to dream and chase after those dreams. All accomplishments start with a dream, the decision to take a first step in what will be a long journey. Thanks Dad! That was a great gift. I have started a lot of long journeys. And if dreams provided the direction in my life, I would argue fear of failure was my biggest motivation in realizing them.

I hope John and my Dad are comfortable in their coffins. I know that's a weird thought. I, for one, have a great fear of being trapped under the ground for eternity. I can't even handle a closed MRI. How would I do if I were to wake up six feet under, alive in my coffin. Hell for me is being trapped in a space in which one cannot turn, cannot itch or clear their nose. Hell is being stuck forever. Hell is the fear that your mind knows the enormity of the crisis but your body can't do a thing about it. Now that's scary. Maybe I should consider the old fashioned bell in the coffin or perhaps just resort to cremation.

I want to share a fear I had as a child and occasionally experience even today. I used to dream I was stuck in bed. Medically I believe this condition is called *sleep paralysis*. Usually it occurs when I sleep face up on the bed. And I could not get unstuck until, with great effort, I would roll myself off the bed. On two occasions I felt I was falling (because I was) and banged my head on the night stand. In Saudi culture they call this *jathume*, i.e. fear of being stuck. And I was amazed that others, even those of different cultures, have experienced the same thing.

Innocent here is my wife, who may someday get in the way of a punch or kick intended for a character or thing in my dream. She is innocent and I certainly don't want to hurt her. These bedtime manners were recently confirmed in a sleep study where I was filmed boxing while sleeping. Recently we purchased a king sized bed and we now have 16 additional inches space between us. So this problem is largely resolved.

But I continued to have crazy nightmare dreams like the time I dreamed my daughter's cat, who hissed at me regularly, locked onto my arm. And in an effort to get the cat off, I scratched myself so badly I opened four deep scratches on my arm (you see I was the cat). I did not know blood was running down my arm. Many of my night problems have been attributed to sleep apnea, however, and I have not had such weird dreams since I began using a CPAP machine.

The movie "Million Dollar Baby" came close to my worst nightmare. The movie's heroine was paralyzed and unable to turn. And she pleaded with her friend to put her out of her misery. And when her friend slipped into her room to administer a lethal injection, she cried a tear of love as he ended her misery.

So I ask my family if I am someday placed in traction with wires through my skull and unable to turn that the doctor "make me more comfortable." That would be most merciful.

Stories of people being trapped beneath the rubble of earthquake debris in Haiti and Mexico City and elsewhere are frightening to me. I view these people dug from the rubble experiencing the closest thing to a worst nightmare one can imagine, especially if they cannot move.

I'm a geologist and most geologists embrace the rational approach that death is just a part of life. We are well versed in death, life, taphonomy, mass mortalities, extinctions, blooms, and the like. Yeah and that makes perfect sense-except when it's your turn. And now at age 61, I can be assured my days remaining on Earth are fewer than the days I have already lived. I suppose that's why I am on this legacy trip. I do give a damn my family and friends remember me and I hope their memories of me will be sweet. But in the end I too will pass the baton, just like everyone else has done from the beginning of time. We came into this world and we will leave it. Of course that doesn't soften the blow. But I kind of liked my time alive and I thank my Mom, my Dad and God for making that possible. I have had a good ride here on Earth and I look forward to Heaven if that comes to be. A life afterward, however, is a tough bridge for the empirical mind.

Babe

Bob's departure shifted the focus of Carolyn's nursing energies to Babe. Babe was no spring chicken when Bob died. She was already 84 years old. And at that ripe old age she decided it was time for a hysterectomy. Upon her recovery from that surgery, Babe developed an exercise regimen she would follow twice a day. "And what was that regimen, Babe?" I asked. "Oh I walk around the table 10 times with my walker" she replied. "Hardly a marathon" was my response! She laughed. "Oh I can still pinch you and make you do what I ask" Babe would say. Despite the apparently easy regimen, Babe became a bit light-headed recently as she lapped the table and Carolyn called for the ambulance. She was given fluids and advised to get more sleep.

Babe has started taking more time for herself, going for an occasional pedicure and a haircut. Her hair occasionally gets so thin and dry that I would remark in Texas twang "Looks like we could start a brush fire." She deserves the best after all these years. And even though she will never be the next *Cover Girl*, we jokingly remark she may someday be the *Cover-Up Girl*. Carolyn takes care of Babe and even plans events for her, like a Mackinac Island Trip to the north end of Lake Michigan and limousine-chauffeured birthday brunches downtown at the Palmer House, the Drake Hotel and the like. It's no wonder we cannot balance the budget.

Babe is a family treasure and the thought of a world without her is too much for the family to endure. I, for one, call her virtually every day regardless of my location. Thank goodness for Vonage while I was overseas. Babe may be frail but her mind is sharper than anyone's in the room.

Babe is the glue that holds the family together. Without her, many of the kids share no special allegiance. In Babe's absence, however we, as siblings, should not fail the basic courtesies learned in grade school, namely please and thank you, wait your turn, and other niceties. There are 12 other people's concerns to consider.

It's always a hoot staying at Babe's. She has absolutely no idea, for example, which items are fresh and which items are expired in her refrigerator. She also has no idea the condition of the basement shower because she has not visited the basement in years. And when you enter the shower you find every shampoo is flea and tick from washing the dogs. That's how it is at Babe's.

Looking Back

Once, when I was in London I decided to go to the Palace Theatre and see the play Les Miserables. I was alone on a business trip. And at one point in the play, when Jean Valjean's courage and love was tested, I sensed the British man sitting next to me was sobbing profusely. I turned and asked if he was all right and he said he was, but added he sobs every time he sees the play. Then I asked how many times he had seen the play and he said this was his 13th viewing. Knowing the price of these tickets were about 125 dollars each, I was astounded and followed with the question, "so what do you do for a living?" He answered "I am a milk man." I knew this was no ordinary play but a very special one that touched all people, regardless of social standing. I have never forgotten that lesson.

The challenge in life, I feel, is to be emotionally touched by our everyday experiences. We need to laugh, listen, and love. This way we know we have lived and can say *we left no song unsung, no wine untasted* (Les Miserables, I Dreamed a Dream).

My wife, Jean has been a great partner for me. I knew the moment I met her in the laundry room at WIU that she was indeed special. She has allowed me to be myself and pursue my studies in becoming a learned person. Additionally she showed my family great generosity by loaning monies when they needed it most. That's the kind of person I admire and the kind of person to whom I am happily married for some 38 years. She has additionally contributed monies to two Indonesian families over the years. I never ask her how much she sends the Indonesians because you can't take a family half way across a raging river and call it help.

I look back and wonder just how it all worked out so well. Certainly we made mistakes. But the past is past and we must move on. We (Jean and I) do care that our legacy will be sweet and we do care that our children will marry partners that will be kind to them.

In my early years, I occasionally suffered from rage. I remember the kids hiding under the beds if I was angry. And I remember the solemn promise I made when they were young that I would never spank them again. I truly love my family and they love me but I had to beat that demon myself. And I can say with certainty I am grateful to have unhinged that wagon.

Unlike the culture in-place when I was in my freshman year (1966) high school in Palos Hills, Illinois, the picture today is considerably more diverse. Back then, my community was largely white and composed of two ethnic groups i.e. the Irish and the Germans, although the Hispanics were increasingly represented as time moved on. The Germans were the builders while the Irish and Hispanics supplied the labor. The centerpiece of their common experience was the Catholic Church and the Catholic Elementary school. Our activities, especially as children, revolved around the church.

I am surprised to hear about so many pervert priests. Yes I accept that abuses happened elsewhere, but I never, in my many years of working in and with the church, experienced such a thing myself. Nor have I heard of my sisters or any fellow altar boys having such an experience. To me, the priests and the nuns were great people that gave up so much in serving. Perhaps the church should get with the program, however. Man was not made to be alone and neither were women. The Bible says this when God created Eve. She was created to be a companion for Adam. For this reason alone, the church should allow priests and nuns to marry. The present system is abnormal and cannot sustain itself.

Post-Retirement (After Saudi)

I retired from Saudi Arabia in September, 2011, but decided I needed to work a few more years. So I initially pursued a job as core sedimentologist. Core is special to geologists because the geologist on-hand is the first person in the world to have ever seen that rock. That is a privilege, to be there for that *grand opening*. It is the same thrill, but arguably on a lesser scale, as the thrill the NASA geologists experienced when they examined the first moon rocks.

A few years back, my family doctor recommended I see a cardiologist to determine if I had signs of heart disease. And after a few minutes on the treadmill, the attending nurse asked if I preferred being called Frederico or Fred? You say what?? No, my name is Gerry. As it turned out, my data was being recorded next door to the file Federico and Frederico's data apparently was recorded to the file Gerry. Sounds harmless eh! Not so. The cardiologist, upon seeing my file, assumed the bad data was mine when in fact, it belonged to Frederico. Two weeks later I was back in the hospital to have an angiogram to see if I needed a stent. Not so! In fact my blockage was determined to be around 15%, normal for a man my age. So what ever became of Frederico? He likely was found dead on the street, released by my cardiologist. I asked my cardiologist for paper copies of the EKG, advising him I was an expert in curve correlation, and he said no, he could not do this. He must have realized he made a mistake. I should have sued. More specifically, Frederico should have sued. After all, he likely was the victim. That's why they call it medical practice, I guess.

I realize I am not the same man in 2013 that I was in high school or college. Recently, I could not find my second shoe despite the fact it was right where it should have been. And I sometimes, although rarely, forget my thoughts in mid-sentence. My wife noticed this and encouraged me to see a neurologist. And over several months, I completed a battery of tests including CT scans, PT scans, MRIs, sleep apnea studies, blood, and cognitive tests.

Initial results revealed a severe shortage of Vitamin B-12 and I began taking weekly injections to aid my short term memory problems. In a related sleep study, I discovered I wake 27 times per hour due to hypoxia-induced snoring (apnea). My first professional diagnosis suggested I had contracted Alzheimer's disease. A second opinion, however, suggested I may just be fine. The scary truth, I am told, is that neurologists only know with certainty if someone contracted Alzheimer's in an autopsy of that person's brain. Comforting!

I still teach college courses at University of Houston Downtown, but I don't, at this time, trust my judgment to return to my role as a professional geologist or even to drive at night. But the good news is my wife Jean has told me she will take care of me, regardless, conjuring memories of the Beatles song *When I'm 64*.

Coming Full Circle

As I complete this book I wonder if I have portrayed a balanced assessment of my life events and interpretations. I tried to get at the real truth, but it may not be possible. Too many years have passed to know *the truth*. We have all re-interpreted events over time. But I can say, with the license of the author, this was how I remember it.

Of considerable concern is a fair assessment of my father. As a young man my father was taught to yield to and obey his father. He otherwise had limited contact with him. His mother, however, revered the very ground on which Bob walked. He was her joy.

I loved my Dad and think of him daily. He gave all he had and gave it selflessly, and for that I totally respect him. And when I think back on my Dad, I see that his tradition of sharing a beer or having a shot of his beloved Crown Royal whiskey with us was simply his attempt to know us more closely, encouraging us to lay down our defenses.

So I end this memoir thanking God for the remarkable people I met along my path and I wish them well. And Dad, if we should meet again in the hereafter, I am convinced I will find you by following the crowd of angels gathering to hear your saxophone rendition of Stranger on the Shore.

Fruitcake Hill and Beyond

This book follows the journeys of the author from his family beginnings on a hill in northern Illinois to numerous postings the world over. The book is both humorous and insightful. I hope you enjoy it as much as I enjoyed writing it.

Author eating a walleye (fish) shore lunch in Manitoba.

About the Author

Gerald Kuecher published Fruitcake Hill in 2008. This sequel documents the author's journey from that simple farmhouse near Chicago to a number of amazing work locations far afield. The author contrasts his risk taking as a Midwestern youth with his life today, focused on his family and the joys of geologic discovery. He also makes numerous observations on the oil business and the differences between foreign workplaces regarding ethics, character, religions, customs, and etiquette. *Fruitcake Hill and Beyond* is a spirited read for those currently practicing or preparing to work in the arena of international exploration, petroleum service, development, the State Department, or other expatriate positions in foreign countries and missions.

Author in Delphi, Greece